How to Knurl

How to Knurl

An Illustrated Guide to Useful and Decorative Work on Metal or Wood

WESLEY E. McCOLGAN

Exposition Press   New York

EXPOSITION PRESS, INC.
50 Jericho Turnpike, Jericho, New York 11753

FIRST EDITION

© 1973 by Wesley E. McColgan. All rights reserved, including the right of reproduction in whole or in part in any form except for short quotations in critical essays and reviews. Manufactured in the United States of America.

SBN 0-682-47550-5

Contents

	Preface	vii
1	Knurls and Knurling	1
2	The Old Method of Knurling	7
3	The Circular Pitch System	16
4	The Diagonal Knurling	34
5	The Multiple Pitch System	49
6	Tools for Knurling	66
	Appendix A: Tables for Working Out Formulas	74
	Appendix B: Tables of Diameters	77

REFERENCE ILLUSTRATIONS

Figure 1 — Knurling Tool Set Up for Straight Knurling	4
Figure 4A — Tool Set Up for Diamond Knurling	14
Figure 18 — Cutting-Off Tool Set Up	69
Figures showing samples of Multiple Pitch Knurling, and examples illustrated by charts	53 to 64

Preface

In presenting my book, *How to Knurl*, I thought that I could contribute a source of valuable information to beginners, and all who do this type of work. What prompted me to get involved in this work was the difficulties I had when I first started to knurl.

I looked through many books on machine shop work and I found very little information on knurling, especially for beginners; so I decided to experiment until I found some of the answers. I have spent a great deal of time preparing this material, and I hope it will be helpful.

How to Knurl

1
Knurls and Knurling

Knurls are referred to as wheels, rolls, and tools, but they are as near to being gears as anything. In the manufacturing of the best grade of knurls, some parts of the formulas and machining operations are practically the same as those used in gear cutting. To demonstrate how closely knurls resemble gears, two straight-tooth knurls of the same pitch can be meshed together as spur gears. And a pair of right and left diagonal knurls can work together as helix gears. But since the profiles of the teeth are designed for a different kind of work they would not work as well as gears, nor will gears do the work of knurls.

TYPES OF KNURLS

There are a wide variety of knurls and pitches; some are made for special jobs. The knurls we are going to deal with are the standard type called circular pitch; they are the most practical for ordinary use. The straight-tooth produces the straight-line pattern on the work. They have a stepping action, since the contact of the teeth with the surface of the metal is slightly interrupted part of the time, causing some vibration. And coarse-pitched, straight-tooth knurls will not work as well on small diameters as diagonals. In this book *pitch* and *TPI* mean teeth per inch.

The left-hand diagonal knurl produces the right-hand pattern on the work, and the right-hand knurl produces the left-hand pattern. Together in pairs they produce the diamond pattern. The contact of the spiral teeth of the diagonals with the metal is more constant and has a smoother rolling action.

Diametral Pitch Knurls. Diametral pitch knurls are engineered to work on standard fractional stock sizes. They cost about twice as much as circular pitch knurls but are well worth the difference for some types of knurling.

Diamond Knurls. In female diamond knurls the points are depressed on the face of the knurls; they produce the male diamond pattern on the work. Male diamond knurls have raised points and produce the female depressed pattern. These knurls are used singly; they cannot do long sections of knurling with the longitudinal feeds. Since the helix angles cross on their face, they are not free to follow the diagonal grooves, like the teeth of two diagonals. These knurls would ruin the job. They could be moved from one setting to several others. They will not work with the multiple pitch system. They will work best with the circular pitch system. They will do fairly well with the old method if they are not too far off track. Other types of knurls are the convex, concave, and conical.

DIMENSIONS OF KNURLS

Recommended dimensions for 8- and 9-inch lathes: O.D. 1/2", width of face 3/16"; O.D. 5/8", width 1/4" or 5/16", either straight or beveled edges. Beveled edges are best for long sections of knurling. The larger the diameter and width of the face the more pressure will be required.

Number of teeth per inch: coarse, 12 to 20; medium 25 to 35; fine 40 to 80. Coarse pitches require more pressure than fine, also more revolutions and passes to finish the job.

THE KNURLING PROCESS

Knurling is accomplished by bringing the knurls into contact with the surface to be knurled, with a considerable amount of pressure; then the machine is started and more pressure is applied. The rotation of the work causes the teeth to cleave their way into the metal and divide it; the pressure and the rolling action forces the displaced metal between the teeth to rise up into crests. The knurling we are dealing with is for round stock only.

As the knurling progresses the V-shaped knurl teeth form the V-shaped grooves or serrations on the surface of the blank. And the V-shaped knurl serrations form the V-shaped ridges, and they may also be called teeth. The finished knurling closely resembles the grooves and teeth on the knurls.

USES

The most common and useful purpose of knurling is to produce an embossed finish on smooth round stock; this provides a good grip for the fingers and hands. When used for decorative purposes light impressions may be sufficient. The knurling process is used extensively in the manufacture of parts for tools, instruments, and machines.

Straight knurling is also useful to key two parts together on the order of a spline. The end, or other parts of a round shaft, can be knurled, then case hardened and pressed or driven into a mating part such as a thumbscrew or wheel. Undersized parts can often be saved by increasing the diameter by straight knurling instead of bushing.

Approximate increase of knurled diameters, with Type 1 and Type 2 knurls—straight, diagonal, and diamond patterns. These increases may vary with different metals; the diamond patterns are usually slightly larger than straight or diagonal.

Standard Type 1 Knurls				*Type 2 Knurls*	
Teeth per inch				*Pitches*	
12	.034	30	.013	14	.030
16	.025	35	.011	21	.022
20	.019	40	.009	33	.012
25	.016	50	.007		
		80	.004		

KNURL-HOLDING TOOLS

Revolving Head. This tool consists of two parts: a shank and a revolving head that holds six knurls. It can be fitted with three pairs of diagonal knurls to produce coarse, medium, and fine diamond patterns, simply by revolving the head to the proper setting; it also makes a half dozen straight-tooth and diagonal patterns.

Self-Centering Head. This is the so-called self-centering tool (Figure 1). It would be more accurate to call it self-equalizing. It consists of a shank and a movable head, which holds two knurls. It is held in place by a pin that extends through a slot cut in the head end of the shank. The pin limits the motion of the head and imparts a floating action. This results in the pressure being equalized on both knurls.

This tool still has to be centered properly by raising or lowering it in the cross-feed tool post before it can be expected to do accurate work. This is the holder we will deal with in this book. It can be raised in the tool post so that a single straight knurl can be used for straight-tooth knurling. Either right-or left-hand diagonal knurling can also be done with this setup by using a single knurl.

Figure 1

To center, adjust the tool so the center line crosses the center of the knurl pin. When using a pair of diagonal knurls for the diamond pattern, the tool should be adjusted so the center line crosses the center of the floating unit and the center of the pivot pin that fastens it to the shank. It may be necessary to replace the tool post ring and the adjusting wedge with washers in order to set the tool low enough to center it properly. A final test can be made by advancing the tool until the knurls press against the blank; then check the centering. The unit must pivot freely against the work.

WORK BLANK MATERIALS

The composition of the different metals has a considerable influence on the performance of the knurls, and may cause some of the double tracking and poor finish. Soft steel was used for all work described herein. The higher the carbon content of steel, the tougher it will be to work; it will require more pressure and should be run at slower spindle speeds and feeds. It is more likely to splinter and break up into chips than soft steel, and it is much more important that the knurls track.

Hardened steel should never be attempted, because the knurls may be ruined. Some hardened steels are still too hard after they are annealed. If the part has to be hard, it is best to use soft steel and do the knurling first and the hardening after.

Brass, aluminum, and other soft metals can be worked faster, and it is easier to make the knurls track. Wood can also be knurled, preferably coarse knurls on hard wood. Cast iron is not suitable for knurling, even if the tracking is good, because it is too brittle.

SPINDLE SPEEDS

Use back gears for all knurling, next to lowest (approximately 70 rpm) for stock up to 1-1/2" diameter, and lowest for stock over 1-1/2".

LONGITUDINAL FEEDS

Coarse pitches should be fed slower than fine. *Coarse:* .006 to .012. *Medium:* .010 to .014. *Fine:* .015 to .018. It may be best to feed diamond knurling a little slower than straight.

CLEANING KNURLS

Knurls should be inspected before using. If there are any chips of metal between the teeth they should be brushed out with an old toothbrush; then clean the teeth and cover them with machine oil. Sometimes during the knurling operation a considerable amount of loose metal collects on the knurls and the work; it may help to do a smoother job if it is brushed off occasionally, without disengaging the knurls.

REENGAGING KNURLS

If for any reason the knurls are disengaged, they can usually be reengaged, providing the imprints are deep enough for the teeth to catch. To do this the knurls are brought up lightly against the surface of the work so that they can be rotated slightly with the fingers until the teeth catch in the grooves. When using two knurls one can be set and then the other. A final check can be made by applying a light pressure on the tool and testing with the fingers. If the knurls do not turn, there is a good chance they are engaged.

When the knurls are far off track, or when coarse knurling is done, especially in the diamond patterns, it may help to use thread-cutting oil on the work.

2
The Old Method of Knurling

It is a good idea to start with straight-line knurling, which is the simplest; only one knurl is used. Place the knurl in the bottom hole of the "self-centering" tool, and move the unit that holds the knurl as far down as it will go. Then place enough large washers over the tool post so that the tool can be raised high enough to be adjusted until the center of the knurl pin is on the center line. Slide the shank as far back in the tool post as it will go; then make the final adjustment.

The centering can be done with either the headstock or tailstock centers, or, better still, by using a centering tool (see chapter 6). It is handy for centering tools and cutter bits in lathe work.

SUPPORTING THE WORK BLANKS IN THE LATHE

Due to the heavy end thrust of the blank toward the headstock during the knurling operation, when supported in the chuck, some provision must be made for preventing the blank from creeping into the chuck. This may be done by center-drilling one end of the blank and turning the other end 1/16" or more—smaller, for a length of 1/2" or so. This will provide a shoulder to press against the jaws. The other end could be mounted on the tail center—a large diameter blank may butt against the face of the chuck—and held by the jaws. Short pieces a little smaller than the diameter of the blank can be cut and placed between the plug and the end of the blank, allowing 1/2" or more for the jaws to grip the work blank.

8 / How to Knurl

If a considerable amount of this type of knurling is to be done, it would be more convenient to make an adjustable plug, as shown in Figure 2. The length could be adjusted with a screwdriver.

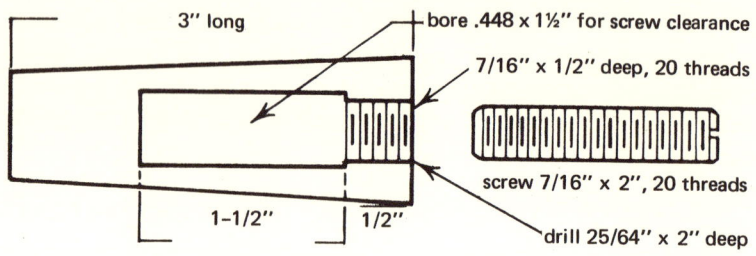

Figure 2. Adjustable Headstock Spindle Plug for a 9" Lathe

Occasionally, it may be more practical to center-drill both ends of the piece to be knurled, mount it between centers, and drive it with a lathe dog. When working on long slender stock it may be necessary to use either a center rest or a follower rest to hold against the side pressure and prevent the work from being sprung out of line or pushed off the centers. When one end of the blank is held in the chuck and the other mounted on the tail center, the extreme side pressure against the tapered side of the center hole causes a strong tendency to push the work off the center. This creates a thrust toward the chuck, also the tailstock. This thrust is greatest when knurling the diamond pattern and using the longitudinal feed. Check tail center to see that it does not work loose. Keep well lubricated.

PREPARING THE SETUPS

Attention should be given to the importance of making every part as rigid as possible. The tailstock ram or spindle should not extend out any farther than necessary.

The dovetail gib adjusting screws on both the compound rest and the cross slide should be tight enough so that the feed screw will not interfere with turning, threading, and cut-off work. The tailstock clamp bolt should be exceptionally tight, also the tool post screw. If the latter is not tight enough the tool may swing sideways and spoil the work when knurling long sections.

After the blank has been properly mounted and the tool centered, the next step is to align the tool at right angles to the blank. Tighten the tool post screw firmly and advance the cross feed until the teeth come in contact with the surface to be knurled. Apply light pressure, rotate the work ahead 1/2" or so, then roll back enough so the tooth marks can be seen. To check for parallelism, compare the imprints with those on the blank in Figure 3, which are described in Chart 1, sections 1 and 2. Oil knurls and surface of blank with machine oil before heavy knurling is started.

Figure 3

THE PRIMARY KNURLING OPERATION

Knurl: 16-TPI straight-tooth, used singly.

Chart 1

Sections	1	2	3	4
Diams.	.975	.975	.975	.965

Section 1. The teeth penetrate deeper on the left side. This indicates the tool is out of line. To correct, tap the heel of the shank toward the right; adjust until the teeth are parallel with blank.

Section 2. The teeth penetrate too deep on the right side; tap the heel of the shank to the left.

Section 3. Shows one light pass made with teeth parallel with blank. The diameters are correct for the pitch of the knurl, so the circumference of sections 1, 2, and 3 contains an exact number of teeth.

Section 4. Shows the teeth begin to step over the initial imprints at the start of the second revolution and strike midway in between. That is as far off track as knurls can get. The teeth are mistracking on this section because the diameter is not right for the pitch of the knurl. Because the diameter is .010 too small, the circumference will not contain an even number of teeth; they start stepping halfway into the spaces. When they are this far off they will require extreme pressure to keep them in mesh. This is the crucial stage in this old method of knurling.

The idea is to sink the knurl teeth deep enough into the metal, about one-third their depth or more, so that by the time the starting point is reached the knurls are actually geared to the work and will not pull out of mesh at the start of the second revolution. At this point the teeth start biting into the extra metal and begin pushing or pulling, and working it to conform to the normal pitch of the teeth.

Due to the extreme pressure required, it is a good idea to start the knurling on the right-hand end of a blank or a shoulder, when practical, and lap the knurls over the edge half their width. This will ease the strain on all working parts and insure proper tracking. Apply plenty of pressure to get sufficient penetration. The work can be rotated ahead, by hand or power, in lowest gear, one revolution. Apply more pressure, then rotate back almost to the starting point; apply more pressure, roll back and forth until the teeth are meshed about one-third their depth. This applies to straight, diagonal, and diamond knurling.

Start lathe, rotate ahead several revolutions, stop and check. The knurls should have the surface of the blank worked into its regular pitch by now. If so, set the longitudinal feed gearing to feed at about .008 toward headstock; then start lathe, engage clutch, and make a test run for 1/4" or more; stop and inspect. If OK, apply plenty of oil to the work part of blank with a small paint brush. Start lathe. When the desired length is reached, disengage clutch, apply more pressure, reverse lead screw, and work toward tailstock.

Repeat this procedure and make as many passes as necessary to bring the crests up to a rather sharp point. Overknurling will cause a poor job, or spoil the work.

If the first test started out bad, check the setup and try again on another section. Sink the teeth in deeper before the starting point is reached. There is a strong tendency for the knurls to pull out of mesh for about one-third of the second round. The reason for this is that most of the extra material is moved and reformed in this area to conform with the pitch of the knurls. The bad starts should roll out fairly smooth by the time the knurling is finished. Often the work can be smoothed up by making the last pass or two without adding pressure. To speed up the work the longitudinal feed may be increased .003 or .004. When knurling long sections, the more the knurls are off track, the more the longitudinal feeds should be reduced.

TEST FOR MISTRACKING

It is good practice to make a test at the start of all knurling to see how far the teeth are out of track. If the blank diameter is too small, the circumference will not contain an even number of teeth, so the knurl teeth will start stepping over the original impressions—if too large they will step under. Either way the effect is practically the same; this is no doubt the cause of most mistracking.

When the tool is set up properly and aligned with the blank, rotate the work ahead one revolution and about 1/4" beyond the starting point; roll back and check the imprints. If by chance the teeth track, they will hold to their tracking even with a very light pressure to start with. But this kind of tracking seldom occurs in this old method, for the simple reason that the diameter is not likely to be right for the pitch of the knurls. No effort is made to calculate the diameter measurements and the pitch of the knurls so that the circumference will contain an even number of teeth.

If the teeth are off .004 or more, it may be a good idea to work the teeth in fairly deep before the starting point is reached. The more the teeth are off track, the more extra material there will be to move, and the deeper the teeth should be meshed before the second revolution is started.

As mentioned before, when at the start of the second revolution the teeth hit midway between the initial imprints; that is as far out of track as the knurls can get. Incidentally, at this point you have a choice of two pitches; a 16-TPI pattern can be produced by using excessive pressure, which is a tough assignment for small lathes; or a 32-pitch can be had by using a light pressure. The trick here is to use the right amount of pressure so the imprints of the second revolution are exactly in the center of the spaces made by the first round, and to feed the tool into the work gradually with the lathe running. This effect can be produced with all standard knurls and pitches. The teeth will divide the spaces into two parts and start an extra set of tracks. This action will multiply the number of ridges and grooves by two: 2 X 16 = 32. And the pattern produced will be practically the same as if a 32-pitch knurl were used. This will be referred to as multiple pitch knurling, which produces a number of patterns of different pitches with the same knurls.

DIAGONAL KNURLING

This is done with a single diagonal knurl either right- or left-hand. The setup and procedure is the same as for straight knurling. The right-hand knurl will produce the left-hand spiral pattern; and the left-hand knurl will give you the right-hand pattern.

DIAMOND KNURLING

Stock knurls for general use are allowed a diametral tolerance of .005" and more by some manufacturers. This may be close enough when knurls are used singly, but when a pair of diagonals are used to produce the diamond pattern, both knurls should be alike in diameter within .001" or less. This measurement will be referred to as "mated"— which means a matched pair. If the diameters of both knurls are within this tolerance they should be well mated and work well together; both should track on the same diameters without using excessive pressure.

Place the pair of knurls to be used in the holding tool; the setup is the same as for straight knurling, except that the centering is different. The tool may be centered with a centering tool, or by bringing it up to the tail center, sliding it ahead and back with the cross-feed screw, and adjusting until the point will pass the center of both the floating unit and the pivot pin. This unit should have free movement so the pressure will be equalized on both knurls when they come in contact with the work. Sometimes these tools will work better when they are set above or below the center line.

If there is much knurling to be done, it is a good idea to place large washers over the tool post and experiment to find the position where the knurls work best; thin shims may also be needed. Then "mike" the thickness and machine a spacer ring the same thickness as this measurement. A ring could also be made for straight knurling. Then when there is knurling to be done, you can use the special rings and forget about centering. That will save time and should be more accurate.

You will now have two knurls to contend with. It is reasonable to assume they will require more pressure than one, and they will be much more complicated and difficult to work with. There may be times when you are puzzled by the erratic behavior of the knurls. There are a number of reasons for this. Some of them will be described as we go along. The cause of some of the difficulties are hard to figure out. You may be amazed to find one knurl tracking true to its pitch and the other making many fine imprints or chewing the metal into splinters. This fault is usually caused by the diameter of one knurl being larger than the other. The best way to eliminate this trouble is to get a pair of the best-grade knurls that are well mated.

This should be done before attempting to correct the other difficulties. Other causes for one or both knurls mistracking, even when well mated, are: they may not be parallel with the work blank, the tools may not be properly centered, one knurl may be duller than the other, or the tools may be old and worn. Some of the difficulties defy solution. In fact, it would be impractical to try to make some things come out perfect.

When using coarse pitch knurls—12 to 20—it may be better to use less pressure and make more passes. The tail center requires extra attention to see that it does not work loose, and it should be kept well lubricated with oil mixed with powdered graphite. Some knurling may need a final pass or two without adding pressure to give the work a smoother finish.

If the knurling tool is set too far above center, it will cause the lower knurl to penetrate deeper than the top knurl. If set too low, the top knurl may penetrate deeper than the bottom knurl. The tool should be set so the teeth penetrate the same depth.

Figure 4A

Figure 4B

Figure 4A shows the tool set up for diamond knurling, and Figure 4B shows an example of diamond knurling, which is analyzed in Chart 2.

Chart 2

Sections	1	2	3	4
Diameters	.561		.594	
Knurls	14-pitch Type 2		20-TPI Type 1	

Section 1. The 14-pitch knurling on this section of the blank was done to show what happens when the knurls are mismated. One knurl was "miked" at .622 diameter; it produced a 14-pitch pattern. The other, at .630 diameter, stepped over the imprints made by the first round, striking midway on the spaces and dividing them into two parts; this caused the knurl to produce a 28-pitch pattern.

When one knurl is .008 larger than the other, there is not much use trying to do a good job with them. It would be best to get a pair of well-mated knurls. It is possible, though, to do a fair job by subtracting .004 from the diameter, which should divide the mistracking between the two knurls, and by lapping the knurls one-half their width over the end of the blank or a shoulder, and sinking the teeth deep into the work before engaging the longitudinal feed, as described for Chart 1, section 4.

Section 2. The 20-TPI knurls were not parallel with the blank; the left side penetrated deeper than the right. The tool post screw should be unscrewed a little and the heel of the shank tapped a little to the right.

Section 3. The light diamond knurling on this section was done with a pair of well-mated 20-TPI type 1 knurls parallel with the blank.

Section 4. The knurls penetrated deeper on the right side. This may be corrected by tapping the heel of the tool to the left.

3
The Circular Pitch System

The circular pitch system of knurling is unlike the old method, where little or no attention is given to the diameters. The circular pitch of the knurls is figured out; that is, the space between the teeth of the knurls and the space between the first impressions of the teeth on the surface of the work blank. One inch equals 1000 thousandths.

For example, the distance measured between the teeth of a 16-pitch straight-tooth knurl would be 1/16". When changed to decimals, as in formula 1, step 1, it is .0625, which is the theoretical circular pitch. The space between the impressions it produces on the work blank is also close to being .0625, but this is affected by the tolerance of the knurls and the arc of the blanks.

If you place a piece of carbon paper on a sheet of white paper, roll a 16-pitch, straight-tooth knurl on it, and measure the imprints, you will find there are 16 spaces to the inch. But there is a difference when the teeth work on a curved surface.

The main principle of this system is to calculate the diameter and circumference so the surface of a blank will contain an exact number of teeth. This permits the teeth to work their way straight into the metal without moving any more material than is necessary. The knurls do not have to be geared to the work in order to force them to hold to their tracking as with the old method.

In straight-line knurling, the number of teeth to be knurled on the work blank times the circular pitch gives you the circumference. Divide this by 3.1416 to get the diameter.

The blanks should be machined true and smooth. This may be done by leaving about .003 for the last cut, and using a fast spindle speed and a slow longitudinal feed. The advantages of this system are: easier starting, less pressure and strain on tools and equipment, and less splintering. It also permits multiple pitch knurling.

DEFINITIONS

Arc means the part of the curved surface of the blank that comes in contact with the knurls. It has an effect on knurling. You can readily see the effect by placing a 1/16" rod between the teeth of a 16-TPI straight-tooth knurl. It will go in between the teeth so deep that it would be impossible for the knurl to work on it. It is not practical to knurl stock much less than 1/4" with a 16-TPI knurl, because it will slip and slide on the surface of the blank.

The larger the diameter the more this curve flattens out, and the less effect it has on the knurling. So the circular pitches of straight knurls and the transverse circular pitches of diagonals should not have to be changed as often with large diameters as with small ones when working out tables of diameters. Diagonal knurls may be used on smaller blanks than straight knurls.

Apparently the interference of the arc in knurling is due to the surface area of the blank increasing between the teeth as the diameters become smaller; the effect is a little more metal between the teeth, and the reverse is the case when the diameters become larger. To correct this effect, the circular pitches and transverse circular pitches can be changed as needed. This can be checked in the tables of diameters (see Appendix), which show that the circular pitches usually become larger as the diameters decrease.

Increment is the difference that occurs in the diameter measurements of the work blanks when one tooth is added to or subtracted from the circumference. It can be used in several ways. For instance, if you have a piece of stock that is smaller than what has been worked out in the tables, you can subtract one or several teeth from the circumference by subtracting the increment from the diameter. Thus, if the smallest diameter shown in the tables of diameters is .228 and the number of teeth on the circumference is 11 and you want one tooth less, you can subtract the increment, which in this example is .020, from the diameter, .228. This would reduce the diameter to .208 and the number of teeth on the circumference to 10.

The diameters can also be increased by adding one or several increments to the diameters without throwing the tracking of the knurls off much.

FORMULAS FOR STRAIGHT-TOOTH KNURLING

Formula 1: Theoretical

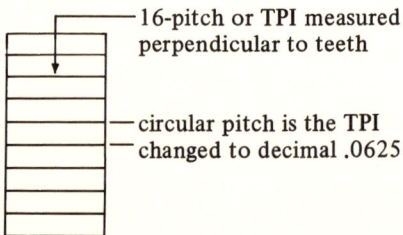

The first and most important thing is to find the circular pitch of the knurls. Formula 1 shows one way it can be done. (A more accurate method is given in formula 3.)

Step 1. Find the circular pitch of the knurl by dividing 1.000 by the TPI. Here is a sample computation for a 16-TPI straight tooth knurl:

```
     16 ) 1.000 ( .0625  circular pitch of knurl
           96
           ――
           40
           32
           ――
           80
           80
```

Thus the circular pitch of this knurl is .0625. To obtain the circumference, multiply the circular pitch by the number of teeth. Then divide the circumference by 3.1416 to obtain the diameter of the blank.

Step 2. To find the increment divide the circular pitch (.0625) by 3.1416. (Use Table 1 in the Appendix for division and subtraction.)

The Circular Pitch System / 19

```
3.1416 ) .0625000000 ( .019894
       31416
       310840
       282744
       280960
       251328
       296320
       282744
       135760
       125664
              terminate
```

Thus the increment for this knurl is between .019 and .020. Use .020.

Formulas 1 and 2 are not recommended for actual work because they are not accurate enough. All of the tables of work blank diameters will therefore be worked out with formulas 3 and 6. There are two reasons for this:

1. Too much tolerance is allowed in the manufacture of some knurls.
2. There is a difference between the arc of a small diameter and a large one. A knurl may track perfectly on a 3/8" diameter blank, and be way off on a 1" diameter. This may be corrected by changing the circular pitches as often as necessary to correspond with the diameters. This feature will be explained and demonstrated later on.

Because there are so many kinds and pitches of knurls and because they are seldom perfect, it is not practical to work out accurate tables of diameters for universal use; only the tables of diameters for a few of the knurls most commonly used will be prepared. If you are fortunate enough to have knurls that correspond with the pitch and measurements described at the head of the tables of diameters, they should track closely enough to permit a good job without using excessive pressure. You may also be able to make use of the multiple pitch system.

Although the circular pitch system is easier and faster, and less pressure is needed at the start, it is important that the job should be completed promptly. Otherwise the rolling action may compact the metal and cause work hardening, which would require more pressure to keep the knurls penetrating.

Formula 2: "Miked"

When the pitch is not stamped on the knurls and you do not know what it is, you can process straight-tooth knurls by using a 1-inch micrometer.

Step 1. "Mike" the knurl by holding it in your fingers so that it can be rotated between the points of the 1-inch micrometer. Rotate several times so that the points barely touch the knurl teeth. Measure carefully several times, so that the measurement will be accurate. For example, suppose you found the diameter to be .628. Next count the number of teeth on the circumference of the knurl. This can be done by dipping the tip of a toothpick in paint and marking a groove between the teeth. Then the thumbnail can be used to count the grooves around the circumference, starting with the marked one and ending with the one next to it. This will give you the number of grooves around the circumference, and there will be the same number of teeth. We will assume you counted 31.

Step 2. To find the circular pitch of the knurl multiply the diameter .628 by 3.1416 to get the circumference and divide the circumference by the number of teeth, which is 31, to get the circular pitch. (Use Table 1 for multiplication.)

$$
\begin{array}{r}
3.1416 \\
.628 \\
\hline
251\ 328 \\
628\ 32\ \ \\
1\ 88496\ \ \ \ \\
\hline
\end{array}
$$

number of teeth 31) 1.9729248 (.0636427 or .06364 circular pitch

$$
\begin{array}{r}
1\ 86\ \ \ \ \ \ \ \ \\
\hline
112\ \ \ \ \ \ \ \\
93\ \ \ \ \ \ \ \\
\hline
199\ \ \ \ \ \\
186\ \ \ \ \ \\
\hline
132\ \ \ \ \\
124\ \ \ \ \\
\hline
84\ \ \ \\
62\ \ \ \\
\hline
228\ \\
217\ \\
\hline
\end{array}
$$

Step 3. To find the pitch or TPI of knurl divide 1.000 by the circular pitch, which is .0636427 for this example:

```
          .0636427 ) 1.000000000 ( 15.71   pitch or TPI of "miked"
circular pitch       636427                knurl; it should pass as
   of knurl          ───────               16 pitch
                     3635730
                     3182135
                     ───────
                     4535950
                     4454989
                     ───────
                      809610
                      636427
                      ──────
                              terminate
```

Step 4. To find the increment divide circular pitch by 3.1416:

```
          circular pitch
3.1416 ) .0636427000 ( .020258   increment for this knurl:
         62832                   variation .020 to .021;
         ─────                   use .020
         81070
         63832
         ──────
         182380
         157080
         ──────
         253000
         251328
         ──────
```

Variation of Increments. The increments are obtained by dividing the circular pitches and transverse circular pitches of all knurls and blank diameters by 3.1416. Since the circular pitches rarely come out in even thousandths, the lesser figures accumulate each time the circular pitches are multiplied by one tooth, like the .0006427 in the circular pitch .0636427, for a 16-TPI knurl. This will cause the diameters to increase .001 at certain intervals. Then the buildup will start over again. This will also effect the increment, causing it to alternate between .020 and .021 at certain intervals.

If all increments came out in even thousandths like .020 and .021, there should be no buildup and no variations. This effect can be tested in the table of diameters for 16-TPI knurls. If you start with the diameter for 50 teeth and subtract the next diameter below it and so on for 10 teeth or more, you find the increment is sometimes .020 and sometimes .021. See page 32 for a way of checking increments.

When the circular pitch of knurls is obtained as in step 2 it can be used to find the diameters and they should be close enough for tests.

Example: We want to knurl 33 teeth on a blank. Multiply the circular pitch (.06364) by the number of teeth (33) to get the circumference and divide the circumference by 3.1416 to get the diameter:

```
              .06364  circular pitch
                  33  number of teeth
              19092
             1 9092
   3.1416 ) 2.1001200 ( .668  diameter of blank
            1 88496
              215160
              188496
              266640
              251328
                       terminate
```

Note: In the table of diameters, the diameter for 33 teeth is .671 and the circular pitch for this diameter is .0639. It was worked out with formula 3, which is more accurate.

Formulas 2 and 3 can be used to process all straight tooth knurls, including the diametral pitches.

Formula 3: "Individual"

Formula 3 is the most accurate we have used so far. It will give you the actual working circular pitches of the knurls. To begin with you need not know anything about the knurls. The principal part of this formula consists of finding the diameter on which the knurls will track perfectly. You then count the number of imprints on the circumference of the blank.

The Circular Pitch System / 23

Knurl: 16-TPI straight-tooth. *Diameter:* .628. *Number of teeth:* 31.

Step 1. Use steps 1 and 2 of formula 2 to find the circular pitch of the knurl you are going to use. For this example we will use a 16-TPI knurl, with a circular pitch of .06364 and an increment of .020258; use .020

Step 2. We will start with a length of 5/8 stock, diameter .625. First we must find the circumference. This is done by multiplying the diameter (.625) by 3.1416. Table 1 can be used for multiplication. Start with the figure 5, then use 2, and finally 6:

```
5 X 3.1416 =     157080
2 X    "    =     62832
6 X    "    =   1 88496
              1.9635000  circumference of stock
```

Step 3. To find the number of teeth the stock will contain, divide the circumference by the circular pitch of the knurl, which in this example is .06364:

```
                    circumference
circular pitch .06364 ) 1.9635000 ( 30.8   number of teeth (use
                        1 9092              30, which is the largest
                        ───────              number of teeth the
                          54300              stock will contain)
                          50012
                        ───────
                           3388  terminate
```

Step 4. To find the circumference that will contain 30 teeth, multiply the circular pitch by 30:

```
  .06364   "miked" circular pitch of knurl
      30   number of teeth to be knurled on blank
─────────
 1.90920   circumference of blank
```

The transverse circular pitch can also be used with steps 2, 3, 4, and 5 when working with diagonal knurls.

24 / How to Knurl

Step 5. To get diameter of blank for the preliminary work divide the circumference (1.90920) by 3.1416. (Use Table 1.)

```
              circumference
    3.1416 ) 1.90920000 ( .607 7    When this figure is over
             1 88496           1    .0005 add .001 to diameter
             ─────────       ─────  diameter for 30 teeth.
               242400         .608
               219912
               ──────
               224880
               219912
               ──────
                         terminate
```

Step 6. To find the diameter on which knurls will track best, see the sketch of blank 1. We can make a rough preliminary test by starting on the right-hand end of blank and turning the full length to be knurled to the largest diameter we have figured out, which is .615. The blanks should always be turned a little larger to start with than the diameter we want, which in this example is .608.

Then reduce each section by .002, as indicated on the blank. This should be faster than .001 in finding the diameter on which the knurls will track. Use just enough pressure so that the imprints show up plainly. Each section can be turned wide enough for at least one test; then disengage the longitudinal feed, take a .002 bite, and move to the next sections.

Blank 1: 30 Teeth

rough diameters   5/8 stock

.605	.607	.609	.611	.613	.615
small	fair	best	fair	large	large

Perhaps the simplest way to find the diameter on which the knurls will track best is to use a piece of stock long enough for 10 or more sections and step down .001.

Step 7. To obtain more accurate measurements we can take the blank used in step 6 and reduce it by one increment, which is .020, as indicated by the sketch of blank 2. Subtract: .615 - .020 = .595, then reduce each section by .001.

Blank 2: 29 Teeth

finished diameters	5/8 stock	.589	.590	.591	.592	.593	.594	.595
		fair	ok	best	ok	fair	large	

Experiment and test until you find the diameter on which the teeth will track perfectly, and will not step over or under the original imprints on the second revolution. Use only enough pressure to make good clear impressions.

In this test we will assume the correct diameter is .591. Mark the end of an imprint with a sharp punch, and start counting from it around the circumference until you come to the imprint next to the marked one. This will give you the number of teeth the circumference will contain. For instance if the number of imprints is 29, the number of teeth produced on the finished work will also be 29. The impressions will penetrate deeper as the knurling progresses; they will depress into grooves, and the spaces between will rise up and form the crests or teeth on the work.

There will be the same number of grooves and teeth. This formula can be used with all straight tooth knurls, including diametral pitch. Sometimes it is a good idea to test the tracking by rotating the work for several revolutions with a little pressure. Then check to see if the knurls are still tracking.

Step 8. To find the circumference multiply the diameter (.591) by 3.1416, using Table 1 for multiplication. Start with first figure of diameter to the right:

```
first figure   1 X 3.1416 =     31416
second  "      9 X   "    =    282744
third   "      5 X   "    =   1.57080
                              1.8566856  circumference of work blank
```

The decimal places may be counted from 3.1416 and the diameter .591.

Step 9. To find the individual circular pitch, divide the circumference by the number of teeth, which in this example is 29:

```
                        circumference
   number of teeth  29 ) 1.8566856 ( .0640236  individual circular pitch
                         1 74            or .064
                           116
                           116
                             068
                              58
                             105
                              87
                             186
                             174
                                    terminate
```

This is the actual working circular pitch of the knurl as related to this diameter. It can be expected to change with different diameters due to the effect of the arc.

After you have obtained the circular pitch of any straight-tooth knurl, it can be used to work out tables of diameters; it need not be more than 5 digits.

Step 10. To find the increment divide the circular pitch by 3.1416, using Table 1.

```
            circular pitch
   3.1416 ) .0640236000 ( .020379   increment for this knurl.
            62832                   Variation .020 to .021;
            119160                  use .020
             94248
            249120
            219912
             292080
             282744
                        terminate
```

The Circular Pitch System / 27

Because of the effect of the arc the increment also changes along with the circular pitches, relative to the changes in the diameters.

Step 11. To find the TPI of the knurl divide 1.000 by the individual circular pitch:

```
.06402364 ) 1.00000000000 ( 15.618   pitch or TPI of knurl as re-
              6402364                lated to the actual work of
              35976360               the knurl. This should pass
              3201182                for a 16-pitch knurl.
              39645400
              38414184
               12312160
                6402364
               59097960
               51218912
```

Step 12. Multiply the circular pitch by the number of teeth to get the circumference, which you divide by 3.1416 to get the diameter:

```
             .064
              25
             320
            1 28
3.1416 ) 1.6000000 ( .509 diameter of blank
         1 57080
           292000
           282744
                 terminate
```

The accuracy of this formula compared with 1 and 2 is demonstrated by the following examples, worked out with a 16-TPI knurl, 25 teeth knurled on blank:

Formula	Circular Pitch	Diameter	Result
1	.0625	.497	too small
2	.06364	.506	too small
3	.064	.509	OK

28 / How to Knurl

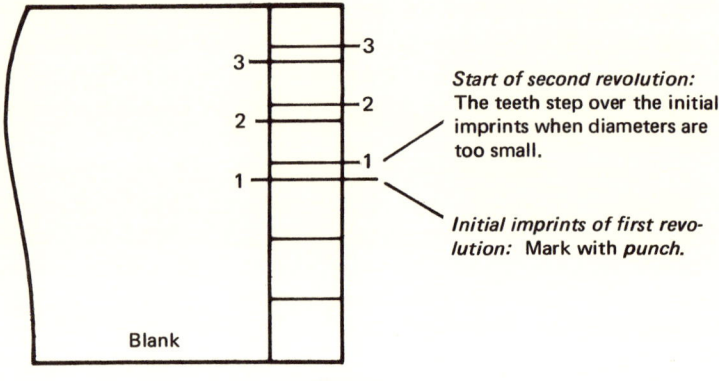

Figure 5

Figures 5 and 6 show how the teeth step over the original imprints when the diameters are too small, and under when too large.

It is apparent that formula 1 is not of much use in working out tables of diameters.

Step 12 can be used with diagonal knurls by multiplying the transverse circular pitches by the number of teeth. We could also work out diameters from 1/4" to several inches if it were not for the effect of the arc interfering with the circular pitches.

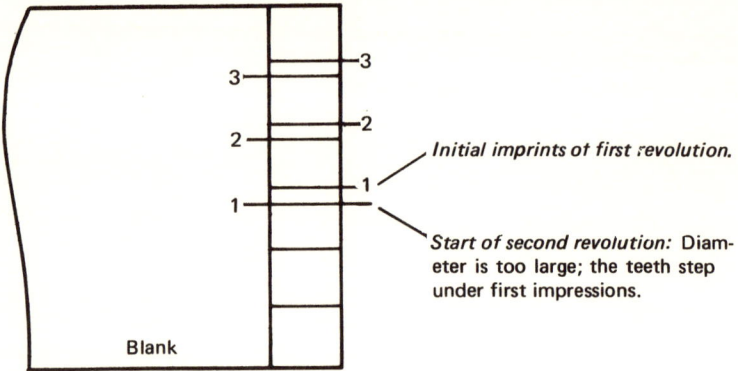

Figure 6

The Circular Pitch System / 29

When using Step 12 to work out tables of diameters it may be best to start with the largest diameter you intend to use, and work down to the smallest. This way if you happen to take off a little more material than you intended, you can reduce the diameter by one increment, which is .020 for a 16-TPI knurl.

For instance, suppose the diameter was to be .407 with 20 teeth on the circumference and you took off .004 too much. You could correct this by subtracting .020 from the diameter: .407 − .020 = .387, which is the next diameter down. This would reduce the teeth on the circumference to 19. This can be done with all knurls by using their individual increments, which can also be added to increase the diameters of the blanks. One increment represents one tooth on the circumference.

The work may be speeded up by using a number of increments. For example, suppose you want to reduce the diameter .387 by four increments:

4 × .020 = .80   .387 − .080 = .307, the diameter to use

When working out tables with step 12, and the diameters become .002 (too small), add .002 or what is needed, and if the diameter is too large subtract the amount needed. When the diameters are this far off it is time to work out a new circular pitch as shown in the two following examples. Refer to Table 5.

To see whether the circular pitch .0644 will work with 11 teeth, we multiply it by 11 to get the circumference, which we divide by 3.1416 to the get diameter:

```
              .0644
                11
              ----
              644
              644
   3.1416 ) .7084000 ( .225
              62832
              -----
              80080
              62832
              -----
             172480
             157080
```

We test this diameter on the lathe. It is too small; the teeth step over the initial imprints. We will therefore have to add several thousandths to the diameter. We can make a test similar to step 7, formula 3, as shown on blank 3.

Blank 3

.226	.227	.228	.229	.230
small	fair	best	fair	large

Now to find the proper circular pitch we can use step 8 and 9, formula 3. Multiply the diameter (.228) by 3.1416 to get the circumference; divide that by the number of teeth (11):

Step 8

```
8 X 3.1416 =   251328
2 X    "    =    62832
2 X    "    =    62832
                .7162848   circumference
```

Step 9

```
11 ) .7162848 ) .0651168   new circumference pitch
     .66         -.0644    old            "            "
     ──          ────────
      56         .0007168  difference
      55
      ──
      12
      11
      ──
      18
      11
      ──
      74
      66
```

There is another way to solve the tracking problem. If we check the circular pitches in the tables of diameters we can see they usually in-

The Circular Pitch System / 31

crease as the diameters become smaller; this is due to the influence of the arc on the circular pitches. When the circular pitches are multiplied by the number of teeth, the circumferences of the blanks will not contain an exact number of teeth. They become too small and at the start of the second revolution the teeth begin to step over the imprints made by the first round. This may be corrected by adding .0001 to the circular pitch and testing the result. If this is not enough, keep adding and testing until the knurls start tracking again. If the diameters become too large the .0001 can be subtracted from the circular pitches.

If you want to knurl a piece of 3/8 stock, formula 3 can be used to find the individual circular pitch and the number of teeth to be knurled on the blank—18 for this example. To find the correct diameter multiply the circular pitch (.064) by 18 to get the circumference and divide that by 3.1416. (Use Table 1.)

```
            .064    individual circular pitch for this diameter:
             18     number of teeth
            512
             64
3.1416 ) 1.15200000 ( .366 6   (anything over .0005 is
          94248            1    added as .001)
          209520         .367   diameter
          188496
          210240
          188496
          217440
          188496
```

Table 2B in the Appendix can be used as an aid to multiplication for determining diameters. It shows the circular pitch .0638 multiplied by factors from 1 to 9. It can be used as in the following example:

Multiply the circular pitch (.0638) by the number of teeth to get the circumference and divide that by 3.1416 (using Table 1) to get the diameter. For this example we will use 39 teeth. Start with 9 × .0638, then 3 × .0638, using Table 2B:

```
                        .0638
                          39
 9 × .0638  =           5742
 3 × .0638  =         1 914
             3.1416 ) 2.48820000 ( .791 9
                      2 19912            1
                        289080         792    diameter
                        282744
                         62360
                         31416
                        309440
                        282744
```

CHECKING INCREMENTS AND DIAMETERS

You can check the increment by subtracting the diameter 1.218 from 1.239. For the 16-TPI the increment should be either .020 or .021. If it is off more than .001, there may be a mistake in the figuring. You should have .020 most of the time for this circular pitch.

Knurl: 16-TPI straight-tooth. Individual *circular pitch for these diameters:* .0638.

```
              .0638  circular pitch
                61   number of teeth
              0638
            3 828
          ) 3.89180000 ( 1.238 7
            3 1416               1
              75020           1.239    diameter for 61 teeth
              62832          -1.218      ”      ”   60  ”
              121880          .021    increment
               94248
              276320
              251328
              249920
              219912
```

The Circular Pitch System / 33

```
         .0638   circular pitch
           60    number of teeth
   ) 3.8280000 ( 1.218 diameter
     3 1416
     ──────
       68640
       62832
       ─────
        58080
        31416
        ─────
        266640
        251328
```

Knurl: 40-TPI straight-tooth. *Individual circular pitch for these diameters:* .02516. *Increment:* .008.

```
                .02516   circular pitch
                  130    number of teeth
                 ─────
                 75480
                 2.516
         3.1416 ) 3.2708000 ( 1.041 diameter
                  3 1416        -1.033    "
                  ──────
                  129200         .008 increment
                  125664
                  ──────
                   35360
                   31416
```

```
    .02516                              .02516
      129                                 128
    ──────                              ──────
    22644                               20128
    5032                                5032
    2 516                               2 516
  ) 3.2456400 ( 1.033 diameter       ) 3.2204800 ( 1.025 diameter
    3 1416       -1.025     "          3 1416
    ──────                              ──────
    104040        .008 increment        78880
     94248                              62832
     ─────                              ──────
     97920                              160480
     94248                              157080
```

4
Diagonal Knurling

OBTAINING PITCHES OF DIAGONAL KNURLS

There are two types of pitches for diagonal knurls in general use: we will refer to them as type 1 and type 2.

Type 1 is based on the normal TPI, and is measured perpendicular to the teeth. This is the standard, used by most manufacturers.

Type 2 is based on the transverse TPI and is measured along the edge of the knurls. This can cause confusion because there is a considerable difference in the patterns. Type 2 will produce a finer pitch pattern. This type seems to have no advantage over the other, because the straight-line knurls and the diagonals of the same pitch do not track on the same diameters and cannot be used on type 1 tables of diameters. If you compare the blank diameters in the tables for the 21-pitch straight knurl and the 21-pitch diagonals, you can see the difference.

The following examples show the difference between the 14-TPI, type 1 and the 14-pitch, type 2 knurls. Formula 1 will be used to work out these problems.

Knurl: 14-TPI, type 1, measured perpendicular to the teeth.

Step 1. To get the normal circular pitch divide 1.000 by the TPI, which is 14:

Diagonal Knurling / 35

```
14 ) 1.000000 ( .71428   normal circular pitch measured
       98                perpendicular to teeth
       ──
        20
        14
        ──
         60
         56
         ──
          40
          28
          ──
          120
          112
          ───
```

Step 2. To find the transverse TPI multiply .86603 by the TPI of the knurls:

```
     .86603   cosine
         14   TPI
     ──────
    3 46412
    8 6603
    ───────
    12.12442   transverse TPI measured along edge of knurls
```

Step 3. To find the transverse circular pitch, divide 1.000 by the transverse TPI:

```
transverse TPI
   12.12442 ) 1.0000000000 ( .082478   transverse circular pitch
              969952                   measured along edge of
              ──────                   knurls.
              300480
              242488
              ──────
              579920
              484976
              ──────
              949440
              848708
              ───────
              1007320
               969952
               ──────
                          terminate
```

The transverse circular pitch is multiplied by the number of teeth to get the circumference, and that is divided by 3.1416 to get the blank diameter to work on.

Knurl: 14-pitch, type 2, measured along edge of knurls.

Step 1. With this type of knurl the transverse TPI is used as the basis for the pitch; thus the 14 is the transverse TPI and you simply divide 1.000 by 14 to get the transverse circular pitch:

$$14 \overline{\smash{)}1.000000} \, (\, .071428 \text{ transverse circular pitch}$$

$$\begin{array}{r} \underline{98} \\ 20 \\ \underline{14} \\ 60 \\ \underline{56} \\ 40 \\ \underline{28} \\ 120 \\ \underline{112} \end{array}$$

The transverse circular pitch is multiplied by the number of teeth to get the circumference, which is divided by 3.1416 to get the blank diameter.

Step 2. To find the normal TPI, measured perpendicular to the teeth, divide the transverse TPI 14, by the cosine (use Table 3 in the Appendix).

$$86603 \overline{\smash{)}14.0000000} \, (\, 16.16$$

$$\begin{array}{r} \underline{8\;6603} \\ 5\;33970 \\ \underline{5\;19618} \\ 143520 \\ \underline{86603} \\ 569170 \\ \underline{519618} \end{array}$$

This is practically the same as a 16-TPI, type 1 knurl.

Diagonal Knurling / 37

Step 3. To get the normal circular pitch, divide 1.000 by the TPI:

```
16 ) 1.0000 ( .0625   the normal circular pitch
       96             for this type knurl
       ──
       40
       32
       ──
       80
       80
```

The two examples below will give you an idea of the difference in the blank diameters between the two types, even though they are designated as the same pitch. We will use 20 teeth to multiply the transverse circular pitch in these two examples in order to find the difference in the diameters:

Type 1
```
              .082478  transverse circular pitch
                   20  number of teeth
table 1 ) 1.6495600 ( .525  diameter of blank
          1 57080
            ──────
            78760
            62832
            ──────
            159280
            157080
```

Type 2
```
              .071428  transverse circular pitch
                   20  number of teeth
        ) 1.42856000 ( .4547
          1 25664         1
            ──────       ───
            171920       .455  diameter of blank
            157080
            148400
            125664
            ──────
            227360
            219912
```

```
Type 1 = .525   diameter of work blank
Type 2 = .455      ”       ”    ”    ”
         ────
         .070   difference in diameters
```

Formula 4: Theoretical (for diagonal knurls)

Knurls: 20-TPI diagonal.

The formulas for diagonal knurls are worked out somewhat differently than those for straight knurls, because the diagonal teeth are cut on a standard 30° helix angle. This is the angle we will use.

Diagonal knurls have two pitches: the normal TPI (measured perpendicular to the teeth) and the transverse TPI (measured along the edge of knurls or knurling). See Figure 7.

The normal circular pitch is simply 20 TPI changed to decimals. For this knurl it is .050. The transverse circular pitch is the transverse TPI, 17.3206, changed to decimals: .0577347. Either right- or left-hand knurls can be used.

Step 1. First we will get the normal circular pitch by dividing 1.000 by the TPI of the knurl, which is 20.

TPI 20) 1.000 (.050   normal circular pitch of knurl
          1 00

Step 2. The cosine of a 30° helix angle is represented by the figure .86603. This must be multiplied by the pitch to get the transverse TPI, before we can find the transverse circular pitch. To find the transverse TPI, multiply .86603 by the normal pitch of the knurls. Use table 3 in the Appendix:

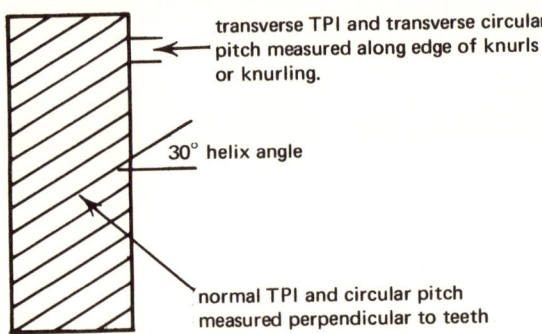

Figure 7

Diagonal Knurling /39

$$\frac{\text{cosine } .86603}{17.32060} \quad \begin{array}{l}\text{normal TPI of knurl} \\ \text{transverse TPI measured along edge of knurl}\end{array}$$

Step 3. To get the transverse circular pitch, divide 1.000 by the transverse TPI:

transverse TPI 17.3206) 1.0000000000 (.057734   transverse circular pitch measured along edge of knurls

```
                866030
                1339700
                1212442
                1272580
                1212442
                 601380
                 519618
                 817620
                 692824
```

Step 4. To find the increment divide the transverse circular pitch by 3.1416, using Table 1:

        transverse
      circular pitch
3.1416) .05773400 (.1837   increment for this knurl
        31416   variation: .018 to .019; use .018
       263180
       251328
       118520
        94248
       243720
       219912

Formula 5: "Miked"

Step 1. Mike the knurls as described in formula 2, step 1, for straight-tooth knurls, to find the diameter and the number of teeth. By applying the magic of figures, we can find out almost anything we want to know about knurls and what they can be made to do.

40 / How to Knurl

For this example we will assume you found the diameter of the knurls to be .625 and the number of teeth on the circumference to be 34. Incidentally, when you count the grooves on the circumference of diagonal knurls, that will give you the number of transverse teeth per inch; thus the calculations will have to be worked out differently with this formula than with formula 2, which is for straight-tooth knurls.

Step 2. To find the circumference of the knurl, multiply the diameter by 3.1416. Then divide the circumference by 34 to get the transverse circular pitch of the knurl:

```
                          3.1416
                        __.625   diameter of knurl
                         157080
                          62832
                        1 88496
  number of teeth  34 ) 1.9635000 ) .05775  transverse circular pitch
  counted on knurl      1 70                 of knurl
                         263
                         238
                          255
                          238
                          170
                          170
```

Step 3. To find the increment, divide the transverse circular pitch by 3.1416, using Table 1:

```
                transverse
                circular pitch
   3.1416 ) .0577500000 ( .018382   increment for this knurl
             31416                  variation: .018 to .019; use. 108
             263340
             251328
              120120
               94248
              258720
              251328
                73920
                62832
```

Step 4. Now we can find the transverse TPI by dividing 1.000 by the transverse circular pitch:

```
transverse
circular pitch .05775 ) 1.00000000 ( 17.316   transverse TPI
                        5775                  of knurl
                        ─────
                        42250
                        40425
                        ─────
                        18250
                        17325
                        ─────
                         9250
                         5775
                         ────
                        34750
                        34650
```

Step 5. To find the normal TPI of the knurl divide the transverse TPI by .86603, using Table 3:

```
                 transverse TPI
cosine .86603 ) 17.31600000 ( 19.994   normal TPI of knurl,
                 8 6603                or 20 pitch
                 ──────
                 8 65570
                 7 79427
                 ───────
                   861430
                   779427
                   ──────
                   820030
                   779427
                   ──────
                   406030
                   346412
```

To find the normal circular pitch divide 1.000 by the TPI, which is 20:

```
20 ) 1.000 ( .050   normal circular pitch of knurl
     1 00
```

Step 6. Check the helix angle of the knurls.

First, you must know the normal TPI. Then find the transverse TPI of the knurls, as shown in step 4. In these examples it is 17.316. This can be divided by the TPI of the knurl, which is 20:

```
             transverse
                TPI
    TPI 20 ) 17.3160 ( .8658
             16 0
              1 31
              1 20
                116
                100          .86603 cosine
                160         -.8658
                160          .00023
```

This should be near enough to indicate a 30° helix angle.

Formulas 5 and 6 can be used to process all diagonal knurls, including the diametral pitches.

Formula 6: "Individual"

Knurls: 20-TPI Diagonal. *Diameter:* .625. *Number of teeth:* 34.

This formula is the most accurate for working out tables of diameters with diagonal knurls. The procedure is basically the same as formula 3 for straight knurling, except that counting the teeth on diagonal knurls will be transverse TPI. This is due to the effect of the helix angle of the teeth, and the result will be that you will have the transverse circular pitch to work with instead of the circular pitch as with straight knurling.

Step 1. First use steps 1 and 2 of formula 5 to find the transverse circular pitch of the knurls that are going to be used.

We will use the same knurls in these examples. The transverse circular pitch will be the same as obtained in formula 5, step 2; it is .05775. We will use a piece of 5/8 cold-rolled stock, diameter .625.

To make the job easier, one diagonal knurl can be used for these tests. If two diagonal knurls are properly mated they should track on the same diameters as a single knurl and produce the diamond pattern. But the knurling tool would have to be centered differently.

Step 2. Find the diameter of the blank to work on.

For this example we will knurl 33 teeth on the blank. Multiply the transverse circular pitch (.05775) by the number of teeth (33), to get the circumference; divide that by 3.1416 to get the diameter, using Table 1:

```
                .05575
                   33
                ─────
                17325
               1 7325
3.1416 ) 1.90575000 ( .6066
         1 88496          1
         ────────       ────
           207900       .607  diameter of blank
           188496
           ──────
           194040
           188496
```

Step 3. Find the diameter the knurls will track on best.

The procedure is virtually the same as in formula 3 except that the transverse circular pitches are used instead of circular pitches.

For this example the diameter is .607; it should be close enough for a rough test (see blank 4).

Blank 4: 33 teeth

diameters	5/8	.603	.605	.607	.609	.611	.613
					small	fair	best

The example shows that .05775, the transverse circular pitch obtained with formula 5, is not accurate enough. The diameter is .006 too small, but it is near enough so that we can work out the "individual" transverse circular pitch with step 4, which is more accurate. Apparently the difference is caused by the effect of the arc.

Use the diameter where the teeth track best. In this example it is .613. Subtract one increment, which is .018 for these knurls:

$$.613 - .018 = .595$$

Finish by turning another series of diameters, stepping down .001 for each section. Figure on having the .595 in about the center of the blank, as illustrated on blank 5.

Blank 5: 32 teeth

diameters	5/8	.592	.593	.594	.595	.596	.597
		small	fair	best	fair	large	

44 / How to Knurl

Step 4: Find the "individual" transverse circular pitch of the knurl. Take the diameter on which the knurl tracks best. In this example it is .594. Punchmark one of the grooves for the starting point and count the imprints. This will give you the number of teeth on the circumference: 32. Next multiply the diameter .594 by 3.1416 to get the circumference and divide that by the number of teeth:

```
4 X 3.1416  =    125664
9 X    "    =    282744
5 X    "    =  1 57080
number of teeth 32 ) 1.8661104 ( .058316  individual transverse
                     1 60                 circular pitch
                     ———                  of knurl
                      266
                      256
                      ———
                      101
                       96
                      ———
                       51
                       32
                      ———
                       19
```

Step 5. To find the increment, divide the transverse circular pitch by 3.1416:

```
            individual trans-
            verse circular pitch
3.1416 ) .05831600000 ( .0185625    increment for this knurl
         31416                      variation: .018 to .019;
         —————                      use .019
         269000
         251328
         ——————
         176720
         157080
         ——————
         196400
         188496
         ——————
          79040
          62832
         ——————
         162080
         157080
         ——————
```

Step 6. To find the transverse TPI divide 1.000 by the transverse circular pitch:

```
individual transverse  .058316 ) 1.0000000000 ( 17.1479  approximate
circular pitch                   58316                    transverse
                                 _____                   TPI of knurls
                                 416840
                                 408212
                                 _____
                                  86280
                                  58316
                                 _____
                                 279640
                                 233264
                                 _____
                                 463760
                                 408212
                                 _____
                                 555480
                                 524844
```

Step 7. To find the normal TPI of knurls divide the transverse TPI by the cosine .86603, using Table 3:

```
              transverse TPI
      .86603 ) 17.1479000 ( 19.802  normal TPI of knurls
               8 6603
               _____
               8 48760
               7 79427
               _____
                 693330
                 692824
                 _____
                 230600
                 173206
```

This should pass for 20 pitch.

To find the normal circular pitch divide 1000 by the TPI, which is 20:

```
        TPI 20 ) 1.000 ( .050  normal circular
                 1.00           pitch of knurls
                 ____
```

Step 8. To check the angle divide the transverse TPI, 17.1497, by the normal TPI, 19.802:

46 / How to Knurl

```
                    transverse TPI
      19.802 ) 17.14970000 ( .86605 approximate cosine
               15 8416
                1 30810
                1 18812
                  119980
                  118812
                    116800
                     99010
```

This should indicate a 30° helix angle. (The cosine of a 30° helix angle is .86603.)

Formulas 3 and 6 are the most accurate for working out tables of diameters. Formulas 2 and 5 are best for processing knurls.

WORKING OUT TABLES OF DIAMETERS

Use formula 3, step 12. The procedure is the same as with straight knurls except that you use the transverse circular pitch instead of the circular pitch. The transverse circular pitch is multiplied by the number of teeth to get the circumference.

Knurls: 30-TPI. *Diameter:* .612. *Number of teeth:* 50. *Increment:* .012.

We will multiply the transverse circular pitch, which is .03883 for these diameters, by 81 teeth and 80 teeth:

```
              .03883  transverse circular pitch
                  81  teeth
               3883
             3 1064
  table 1 ) 3.1452300 ( 1.001  diameter
             3 1416       .989   "        80 teeth
               36300      .012  increment
               31416
```

```
                    .0 3883
                    80 teeth
        table 1 ) 3 10640000 ( .988 7
                   2 82744         1
                    278960       .989  diameter
                    251328
                    249920
                    219912
```

Knurls: 20-TPI. *Transverse circular pitch for these diameters:*
.0589.

```
                     .0589
                     20 teeth
        table 1 ) 1.17800000 ( .374 9
                    94248         1
                    235520      .375  diameter
                    219912     -.356    "
                    156080      .019  increment
                    125664
                    304160
                    282744


                     .0589
                     19 teeth
                     5301
                      589
        table 1 ) 1.1191000 ( .356 diameter
                    94248
                    176620
                    157080
                    195400
                    188496
```

Note: 20-TPI knurls with the same measurements as those used with these examples should track on 3/8 stock, even with the old method. It is seldom that knurls will track well on standard stock sizes, with light pressure.

48 / How to Knurl

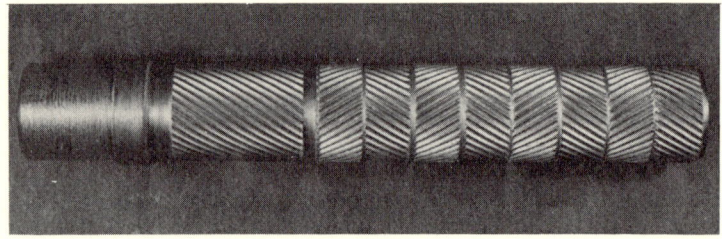

Figure 8

All standard knurls, including the diametral pitches, can be used with formula 3, step 12. For straight knurling, multiply the circular pitches by the number of teeth. For diagonal and diamond knurling, multiply the transverse circular pitches.

When the tool is centered properly both diagonal and diamond knurling should work on the same diameters.

The sample of knurling shown in Figure 8 was done with a left- and a right-hand, used alternately, with the same setup as for straight knurling. A left-hand knurl used alone produces the right-hand diagonal pattern; the right-hand knurl produces the left-hand pattern. (See Chart 3.)

Blank: 36 teeth. *Diameter:* .535, taken from tables of diameters.

Chart 3

	section 1				section 2			
blank diameter .535								

patterns	RH	LH	RH
pitch	25	25	
longitudinal feed	.016	stationary	

Section 1. Right-hand pattern done with a left-hand knurl.

Section 2. The knurling on this section was made by alternating the right- and left-hand knurls. Four settings were finished with a left-hand knurl, leaving spaces wide enough in between for four settings of right-hand.

This type of knurling produces a herringbone pattern.

5
The Multiple Pitch System

"Multiple pitch" means that a variety of pitches can be obtained with one knurl for straight and diagonal knurling, and with two knurls for diamond knurling. In this work you need calculations and careful measurements of the blank diameters. Also you must exercise an exact control over the amount of pressure applied to the knurls.

In the regular circular pitch system we try to prevent the teeth from mistracking. In the multiple pitch system we *want* the teeth to mistrack, but under control. Thus the space between the imprints made by the first revolution on the work blank, which is the regular pitch of the knurls, can be divided into a variety of equal parts.

For example, a 16-TPI straight-tooth knurl can produce the following patterns of pitches: 16, 32, 48, 64, 80, and even finer. Pitches finer than 80 are seldom used.

The patterns transferred from the knurls to the blank would approximate the pattern produced by conventional knurling. For instance, it would be hard to tell a 32-pitch pattern from one made with a 32-TPI knurl.

The included tooth angle of standard knurls is 90° from 12-TPI knurls to 50-TPI, and 70° from 50- to 80-TPI. This tooth angle of the finer pitches results in the teeth being thinner and sharper. They will do a better job than the multiple pitch knurling done with coarse knurls, which should not be expected to be perfect. They are also much more difficult to work with, but they will do fairly well in a pinch.

It is advisable to start practicing and experimenting with any coarse knurl by multiplying the TPI by 2; that would mean the spaces between the original imprints would be divided into 2 parts.

This type of knurling is not intended for production work. It can be useful for occasional knurling jobs in small shops and home workshops where there are only a few knurls on hand. Besides the expense, it sometimes takes days or weeks to have knurls delivered.

DEFINITIONS

Pressure is the force applied to the knurls by advancing the cross-feed screw. The amount in thousandths may be measured by the figures and graduations on the micrometer collar of the cross-feed screw. This is by no means perfect but should be more accurate than guessing.

The pressure on the knurls and work blank can be applied by advancing the cross-feed screw. The micrometer collar has a graduating scale cut on the circumference. Each line indicates one thousandth of an inch. Each ten-thousandth is numbered 0-10-20, up to 90. It can be used to measure the amount of pressure required.

The compound rest screw works independently of the cross-feed screw; it permits the cross-feed screw to be set on zero and the compound rest screw used to take up the slack.

Prepare the blank by turning it to the proper diameter for the pitch selected. Advance the cross-feed until the knurls are close to the blank. Set the reading on the collar at 0. Then advance the compound rest feed screw so the knurls touch the blank. Apply a small amount of pressure on the cross-feed.

Experiment until the correct pressure is found to divide the spaces as desired. It is best to divide the spaces into two parts to start with. Then check the reading on the collar.

Now move the tool to another part of the blank and try for a finished job of full knurling, by using the same amount of pressure as noted on the collar reading. Rotate blank ahead 1-1/4 revolution. Back up enough so the imprints of the second revolution can be seen.

If the imprints of the teeth strike a little below the center of the spaces, use more pressure; if above, use less. Rotate back to starting point and switch on power. Feed the tool into the work slowly at first.

The spaces should come out evenly divided as in Figure 9, section 2, even though several settings turned out bad. If the impressions are not too deep they will roll out to a good finish.

Saturate the work with oil and set the gearing for longitudinal feed of about .015. Start the lathe and advance the tool slowly several thousandths. Engage the clutch; the travel can be in either direction.

When the desired length of knurling is reached, release the clutch, advance the tool, reverse the longitudinal feed, and engage the clutch. Make as many passes as necessary to bring the knurling to a full crest.

As soon as the knurls are found to be tracking properly, a considerable amount of pressure can be used on the coarser pitches, and a very light pressure for fine pitches. Too much pressure will cause the knurls to work into their regular pitch.

Cycle means that a certain number of revolutions are required to fill the spaces with impressions. If the diameter and pressure are right, the spaces will be divided equally.

For example, when the spaces are divided by 2, two revolutions will fill the spaces. Divided by 3, three revolutions will fill the spaces and complete the cycles. When the cycles are completed the teeth will start retracking and working in the same impressions until the job is finished.

Graduations are the edges of the impressions that are made to extend out beyond the main part of the knurling. The purpose of this is to make the imprints easier to count when you ascertain the pitch of the pattern.

After the spaces have been evenly divided, the graduations are made by setting the longitudinal feed at .015 to .020 and engaging the clutch. The lathe is then started; the travel may be either right or left, depending on which edge of the knurling is to have the graduations.

As the knurling progresses the edges of the teeth will advance, causing the top imprints to move ahead of the ones below. If you want the graduations to remain, stop the lathe, release the clutch, and without releasing the pressure move the carriage so that the knurls will slide back into the knurled part about 1/8". Apply pressure; then the longitudinal feed may be changed to the most efficient feed, and reversed so that the travel is to the right. Release the clutch and stop each pass about 1/8" from the end where the graduations are, so that they will not be rolled out.

This procedure will work with all standard straight and diagonal knurls, and with diametral pitch knurls. When working on the diamond patterns, it may be necessary to release some of the pressure in order to slide the knurls back into the knurling. Male and female diamond knurls cannot be used with the longitudinal feeds or the multiple pitch system.

Another way the graduations can be retained on the work when making the finer pitches is to stop the lathe about 1/8" short of the desired length of the knurling. When the last pass is made, the knurling can be finished the full length, and the lathe shut off before the longitudinal feed clutch is released and the tool withdrawn.

The graduations are useful at the start of multiple pitch knurling. After the spaces are properly divided and the imprints are deep enough to be plainly seen, the longitudinal feed can be set at .016 to .020 and the knurling extended for 1/8" or more. Then the lathe is stopped without releasing the longitudinal feed clutch. The imprints should be easy to count. Then the lathe can be started and the work finished.

To check the pitches start with the shortest imprint along the edge of the knurling and count the imprints, including the longest. See Chart 4, section 4. There are four imprints on the spaces between the normal 16-pitch impressions. This means that the 16 pitch is multiplied by 4 and the multiple pitch is 64.

At the start of multiple knurling the blanks should be turned to the diameter where the knurls will track with a light pressure, as indicated in section 1 of Chart 4. When the figures obtained by dividing the increments are subtracted from the blank diameters, the multiple pitch knurling should be more accurate.

EXAMPLES OF MULTIPLE PITCH KNURLING

The five patterns in figure 9 were all made with a 16-TPI knurl. The preliminary straight knurling shown in figures 9, 10, and 11 was done with light impressions so that the spaces can be seen more easily. The blank was divided into five sections as shown in Chart 4. The diameter of each was calculated to produce the different pitches.

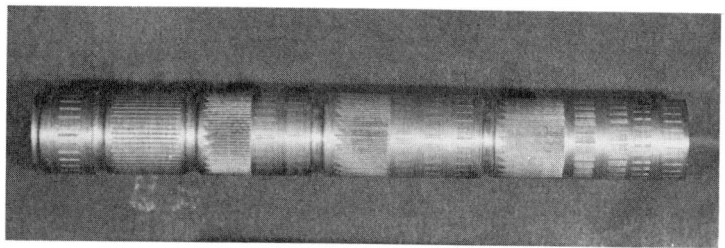

Figure 9

Knurl: 16-TPI straight-tooth. *Diameter:* .628. *Number of teeth:* 31. *Circular pitch of knurl:* .063643. *Increment:* .020258. *Variation:* .020 to .021; use .020.
Blank: 28 teeth. *Diameter:* .570 (taken from tables of diameters.) *Circular pitch for this diameter:* .064.

Chart 4

sections	1	2	3	4	5
blank diameters	.570	.560	.563	.565	.566
pressure	.025	.020	.020	.015	.015
revolutions	1	2	3	4	5
pitches	16	32	48	64	80
longitudinal feeds	.014	.015	.017	.017	.017
teeth	28	54	82	110	138

To start multiple pitch knurling, divide the increment of the knurl, .020, by 2, 3, 4, and 5 to find the amount to subtract from the diameter of the blank, which is .570. This should bring the diameters near enough for tests.

54 / How to Knurl

Increment Table

Increment					Blank diameter			Multiple pitch diameters	Sections		Pitches
					.570				1	=	normal
.020	÷	2	=	.010	→ .570	- .010	=	.560	2	=	multiple
.020	÷	3	=	.007	→ .570	- .007	=	.563	3	=	"
.020	÷	4	=	.005	→ .570	- .005	=	.565	4	=	"
.020	÷	5	=	.004	→ .570	- .004	=	.566	5	=	"

These measurements should bring the diameters near enough so that final adjustments may be made by applying the right amount of pressure.

The work on this blank was started on the right-hand end of section 5. One revolution was taken, showing the regular pitch of the knurl. Then the tool was withdrawn and moved to the left for 2 revolutions, then moved again for 3, then for 4.

The fifth setting shows 5 revolutions, which filled the spaces with impressions. The longitudinal feed was engaged and one pass was taken to the left: 5 X 16 = 80, which is the pitch produced on this diameter. This procedure was repeated on section 4. It took 4 revolutions to complete the cycle: 4 X 16 = 64, the pitch for this diameter.

On section 3 the cycle was completed with 3 revolutions: 3 X 16 = 48 pitch. Section 2 the cycle was finished with 2 revolutions: 2 X 16 = 32 pitch. Section 1 is the normal pitch of the knurl.

This was done to show how each succeeding revolution divides the spaces and multiplies the impressions until the spaces are filled and the cycles completed. Then the teeth step over into the next space and start retracking and forming the patterns on the blanks.

The spindle was rotated by hand until the spaces were filled with imprints. Then the longitudinal feed gears were set at the figures indicated under each section. These feeds are for the long sections of knurling done on the last setting of sections 5, 4, 3, and 2. Then the lathe was started, and the light knurling extended for a short length. A good magnifying glass should be used to check on the finer pitches.

As the knurling progresses the imprints develop into grooves, and the spaces between them form the teeth on the blank. When working with the fine pitches we must be careful and avoid applying too much pressure, because this will have a tendency to cause the teeth to step over a little too far. If too much pressure is used, the knurls may be forced into a coarser pitch or even their regular pitch. The finer the pitches the more gradually the knurls should be fed into the work.

The figures obtained by dividing the increments to get the desired multiple pitches may be added to the diameters. The result should be practically the same as when subtracted. One advantage would be that larger blank diameters could be used in some cases, and less material would need to be turned from the blank.

For example, a blank could be turned .010 larger than in Chart 4, section 2. This should cause the teeth to step under the imprints made by the first revolution and strike near the center of the spaces, dividing them into 2 parts. This would produce the 32-pitch pattern, the same as section 2, and the blank diameter would be .580, or .020 larger than the .560 on section 2. In these examples, the figures obtained by dividing the increments will be subtracted from the blank diameters.

The sample in Figure 10 shows what happens when the diameters are too large or too small. The increment .020 was divided by 3 in order to make the 48-pitch pattern: .020 ÷ 3 = .0066; use .007. This was subtracted from the blank diameter.

Knurl: 16-TPI straight-tooth. *Diameter:* .628. *Circular pitch of knurl:* .063643. *Increment:* .020258. *Variation:* .020 to .021; use .020. *Blank:* 30 teeth. *Diameter:* .610 (taken from the tables of

Figure 10

diameters): .610 - .007 = .603, the diameter to work on. The diameters obtained in this manner should be near enough for tests, provided the right amount of pressure is used.

In sections 2, 3, and 4 of Chart 5 the initial round was completed. Then the pressure was released, the tool moved to the left for the second setting and 2 revolutions, then to the third setting and 3 revolutions. The first setting is the regular pitch of the knurl.

Chart 5

sections	1	2	3	4
blank diameters	.610	.606	.603	.600
normal pitch	16		48	32
pressure	.020	.015	.015	.015
revolutions	1		3-2-1	2-2-1
pitches	16		48	

too small for the 16-pitch

Section 1. The diameter is correct for the normal pitch of the knurl.

Section 2. Was turned .004 undersized. This caused the knurl teeth to step over the first imprints. This interfered with the normal pitch; 3 revolutions did not fill the spaces, but one more revolution may have produced a 48-pitch pattern.

Section 3. The diameter is correct for the 48-pitch pattern; the spaces are divided evenly.

Section 4. Was turned .010 undersized. The teeth stepped over the initial imprints too far. They strike midway in the spaces on the second revolution. This will cause the teeth to retrack on the third revolution and produce a 32-pitch pattern instead of a 16.

This sample of knurling shows that knurls hold to their normal pitches during the first revolution up to the first imprint. If the diameter is too small the teeth will begin to step over the initial imprints; if the diameter is too large the teeth will step under the initial imprints. In either case the normal pitch is disrupted. If too large it can be corrected by filing or turning a very light cut on the blank; if too small the diameter can be reduced by 1 increment of the knurl; for the 16-pitch this is .020. Because the diameters of sections 2, 3, and 4 are too small

The Multiple Pitch System / 57

for good tracking, the same piece of stock can be used for further tests, by subtracting .020 from .610 = .590.

Figure 11 shows that pressure alone in varying amounts can create different patterns on the same diameters. The trick is to get the spaces divided evenly.

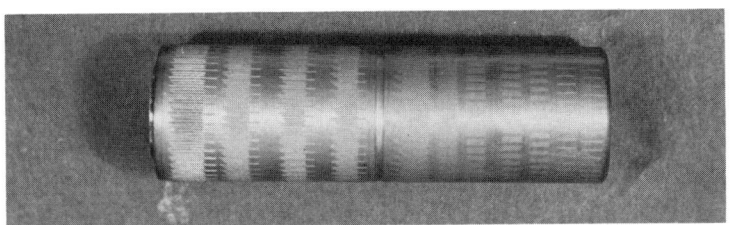

Figure 11

Knurl: 16-TPI straight-tooth. *Diameter:* .628. *Circular pitch of knurl:* .063643. *Increment:* .020258. *Variation:* .020 to .021; use. 020.

Blank: 47 teeth. *Diameter:* .954. This was taken from the tables of diameters and turned for an 80-pitch pattern. To find the amount to subtract from the blank, divide the increment: .020 ÷ 5 = .004; .954 − .004 = .950.

Chart 6 (in two parts)

sections	1	2	3	4			
blank diameter .950							
revolutions	3	4	4	5			
pressure	.036	.030	.026	.020			
pitches	48	64		80			

sections				5			
blank diameter .950							
revolutions	7	6	5	4	3	2	1
pressure	.012						
pitches	112						

Section 5 shows how the spaces between the original imprints divide and increase with each succeeding revolution. Each revolution adds approximately 16 more imprints to the inch. The 7 revolutions fill the spaces:

7 X 16 = 112 the approximate pitch of the pattern

Section 4. 5 X 16 = 80.

Section 3 shows that the spaces were not divided evenly. Thus .026 was not enough pressure; too much gap was left at the top of the spaces on the last round.

Section 2. When .030 pressure was applied the spaces were divided evenly: 4 X 16 = 64 pitch.

Section 1. 3 X 16 = 48 pitch.

If the spaces do not divide evenly, the difficulty can sometimes be corrected. For instance if the last revolution of the cycle leaves a little too much space, as in section 3, the spindle can be rotated back to the starting point and a little more pressure applied. If the teeth step over too far a little pressure can be released, and the power switched on for a test.

STRAIGHT-TOOTH PATTERNS

Figure 12 shows a finished job in which the knurling is brought to a full crest on 5 sections, each with different diameters and pitches. All were done with one 16-TPI knurl. Chart 7 explains how to do this type of knurling.

Figure 12

Knurl: 16-TPI. *Circular pitch:* .063643. *Increment:* .020.
Blank: 41 teeth. *Diameter:* .832 (from table of diameters).

Chart 7

sections	1	2	3	4	5
blank diameters	.832	.822	.825	.827	.828
pressure	.022	.020	.017	.015	.015
revolutions	1	2	3	4	5
pitches	16	32	48	64	80
longitudinal feeds	.013	.014	.015	.015	.015
teeth	41	81	122	163	200

To produce the desired pitches, you must obtain the correct diameters. First find the blank diameters as listed in the tables. Then divide the increment .020 by 2, 3, 4, and 5. Subtract the result from the blank diameter, which in this example is .832:

Increment Table

Increment				Blank diameter			Multiple pitch diameters	Sections	Pitches
				.832				1 = 16	normal
.020 ÷ 2 =	.010	→	.832 −	.010	=	.822		2 = 32	multiple
.020 ÷ 3 =	.0066	→	.832 −	.007	=	.825		3 = 48	"
.020 ÷ 4 =	.005	→	.832 −	.005	=	.827		4 = 64	"
.020 ÷ 5 =	.004	→	.832 −	.004	=	.828		5 = 80	"

SQUARE PATTERN

The square patterns shown in Figure 13 were produced by cutting across the straight patterns shown in Figure 12 with a thread-cutting tool, but on different diameters.
Knurl: 16-TPI. *Circular pitch:* .063643. *Increment:* .020.
Blank: 36 teeth. *Diameter:* .732 (taken from tables of diameters).

60 / How to Knurl

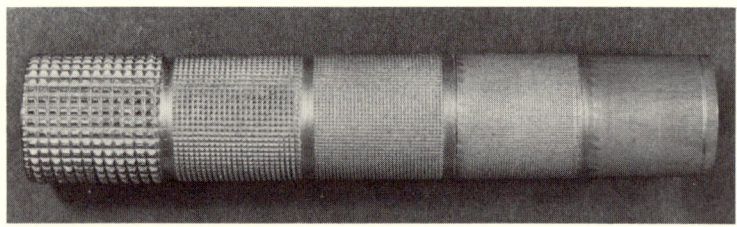

Figure 13

Chart 8

sections	1	2	3	4	5
blank diameters	.732	.722	.725	.727	.728
pressure	.020	.020	.019	.017	.015
revolutions	1	2	3	4	5
pitches	16	32	48	64	80
longitudinal feeds	.013	.014	.015	.015	.015
thread pitches	16	32	48	64	80

Increment Table

Increment				Blank diameter		Multiple pitch diameters		Sections	Pitches
				.732				1 = 16	normal
.020	÷ 2	=	.010	→ .732	− .010	=	.722	2 = 32	multiple
.020	÷ 3	=	.0066	→ .732	− .007	=	.725	3 = 48	”
.020	÷ 4	=	.005	→ .732	− .005	=	.727	4 = 64	”
.020	÷ 5	=	.004	→ .732	− .004	=	.728	5 = 80	”

When finished with the straight knurling, set up for threading. If you use a 16-TPI knurl, any or all of the patterns can be made as follows:

For the 16-pitch, set the gearing for 16 threads per inch. The pitches of the threads and knurls are identical in this example, but the pitches of standard threads may not always correspond with the pitch of the knurls;

in that case the nearest thread pitch may be used. The thread pitches should be the same as the knurled patterns, as indicated in each section. The threading is done as usual. Thread-cutting oil should be applied generously throughout the threading operation. A small paintbrush is ideal for this purpose. Stop the lathe occasionally to check; the ridges should be square, and the threading should be about the same depth as the knurling. The 5 straight and 5 square patterns were produced with one knurl, 10 patterns in all. A still wider variety of rectangular patterns can be made by using different thread pitches on the same straight knurling.

DIAMETRAL PITCH

The knurling in Figure 14 shows that diametral pitch knurls can be processed with formula 3 and used in circular pitch knurling, as well as in multiple pitch knurling, as shown in Chart 9. Besides their special diametral pitch, these knurls can be forced to work with the old method, like any other knurls, where no consideration is given to the diameters. This procedure includes diagonal and diamond knurling.

Figure 14

Knurl: 96 DP, straight-tooth, 30.6-TPI. *Diameter:* .621. *Number of teeth:* 60. *Circular pitch of knurl:* .03273. *Increment:* .01042. *Variation:* .010 to .011; use .010.
  Blank: 60 teeth. *Diameter:* .625. *Circular pitch for this diameter:* .03273.
  Knurl: 30.6-TPI or 31-pitch.

Chart 9

sections	1	2	3
diameters	.625	.620	.622
pressure	.030	.025	.020
revolutions	1	2	3
pitches	31	62	93
longitudinal feeds	.016	.016	.014
teeth on blank	60	119	179

To find the diameters divide the increment .010 by 2 and 3:

Increment			Blank diameter	Multiple pitch diameters	Sections	Pitches
			.625		1 = 31	normal
.010 ÷ 2 =	.005	→	.625 - .005 =	.620	2 = 62	multiple
.010 ÷ 3 =	.0033	→	.625 - .003 =	.622	3 = 93	"

The standard diametral pitch knurls work surprisingly well within their range, which is up to about one inch; beyond that the effect of the arc begins to interfere with the tracking. But they can be used far beyond this point with the above methods

The 64 and 128 diametral pitches are engineered to track on blank diameters having fractional increments of 1/64". The 96 and 160 diametral pitches are for blank diameters having fractional increments of 1/32".

MULTIPLE PITCH DIAGONAL KNURLING

The knurling in Figure 15 shows that multiple pitch knurling can also be done with single diagonal knurls.

Knurl: 20-TPI. *Diameter:* .625. *Number of teeth:* 34. *Transverse circular pitch of knurls:* .05775. *Increment:* .01838. *Variation:* .018 to .019; use .018.

Blank: 33 teeth. *Diameter:* .613. *Circular pitch for this diameter:* .05832 (from the tables of diameters).

The Multiple Pitch System / 63

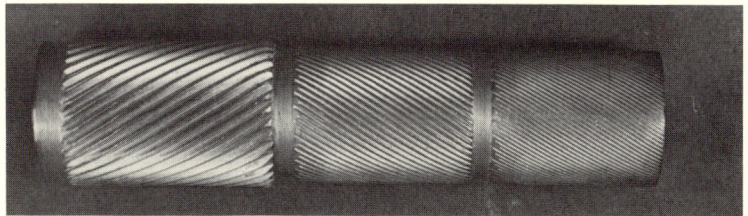

Figure 15

Chart 10

sections	1	2	3
diameters	.613	.604	.607
diagonal patterns	LH	RH	RH
pressure	.020	.016	.015
revolutions	1	2	3
longitudinal feeds	.015	.014	.014
pitches	20	40	60

In order to work out the desired pitches, the increment .018 was divided as shown below:

Increment	Diameter	Multiple pitch diameters	Sections	Pitches
	.613		1 = 20	normal
.018 ÷ 2 = .009 → .613 − .009 =		.604	2 = 40	multiple
.018 ÷ 3 = .006 → .613 − .006 =		.607	3 = 60	"

At the start of all multiple pitch knurling the blanks should be turned to the diameter where the knurls will track with a light pressure; this is especially important when working on the fine pitches. Then when the figures obtained by dividing the increments are subtracted from the blank diameters the multiple pitch knurling should be more accurate.

MULTIPLE PITCH DIAMOND KNURLING

The knurling in Figure 16 was done with a pair of 20-TPI knurls; the right-hand was .6255 diameter and the left-hand .625 diameter. A pair of knurls with a tolerance this close may be considered well mated.

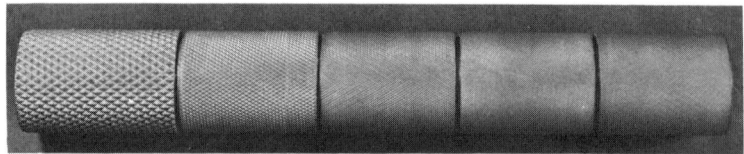

Figure 16

Knurls: 20 = TPI. *Diameter:* .625. *Number of teeth:* 34. *Transverse circular pitch of knurls:* .05775. *Increment:* .01838. *Variation:* .018 to .019; use 018.

Blank: 41 teeth. *Diameter:* 760. *Transverse circular pitch for this diameter:* .05815 (from the tables of diameters).

Chart 11

sections	1	2	3	4	5
diameters	.760	.751	.754	.755	.756
pressure	.020	.015	.007	.006	.006
revolutions	1	2	3	4	5
longitudinal feeds	.012	.013	.014	.015	.015
pitches	20	40	60	80	100

To find the amount to subtract from the blank diameters, divide the increment .018 by 2, 3, 4 and 5 as shown below:

Increment					Blank diameter				Multiple pitches diameters	Sections	Pitches
					.760					1 = 20	normal
.018	÷	2	=	.009	→ .760	−	.009	=	.751	2 = 40	multiple
.018	÷	3	=	.006	→ .760	−	.006	=	.754	3 = 60	"
.018	÷	4	=	.0045	→ .760	−	.005	=	.755	4 = 80	"
.018	÷	5	=	.0036	→ .760	−	.004	=	.756	5 = 100	"

These measurements should bring the diameters near enough so that the final adjustments may be accomplished by applying the right amount of pressure.

HOW TO CORRECT MISTRACKING

To begin with we must realize that diamond knurling is complicated and difficult. Everything must be just about right when attempting to knurl the finer pitches. Even when a tool-post spacer ring has been made to center the tool for diamond knurling, it may do well with coarse pitches, but when it comes to the 60, 80, and 100 with the light pressures required, the knurls may mistrack.

For example, we are knurling the 60-pitch pattern as on section 3. The top knurl is dividing the spaces into 3 equal parts; the bottom knurl is rolling 80-TPI more or less, or perhaps chewing the metal into splinters. This would indicate that the teeth are either stepping over or under the imprints made by the previous revolution. If the teeth are stepping under a few thousandths, this may be due to the bottom knurl's being a little smaller than the top one. It may not be as sharp, or the pressure may be lighter. In either case this difficulty may be remedied by placing thin shims crosswise under the head end of the tool shank; this should raise the tool and cause a little more pressure to be applied to the bottom knurl, which should cause the lower knurl teeth to step over farther and penetrate deeper.

If the teeth step over too far the shims should be inserted under the heel end of shank. If the top knurl is at fault and the teeth step under the previous imprints, the shims may be placed under the heel end of the shank, to make it step over more and penetrate deeper. If it steps over too far, shims under the head end of tool may correct the mistracking.

It may require a considerable amount of patience and experimenting to make these fine adjustments.

It is possible to produce 15 different patterns with a pair of 20-TPI knurls: 5 left-hand diagonals, 5 right-hand, and 5 diamond. There are three patterns on the same diameters.

Diamond knurling should work on the same diameters as diagonal knurling. Working on the fine pitches above 60 is very difficult with this system, but it can be done.

6
Tools for Knurling

CENTERING TOOL FOR A 9" LATHE

The centering tool is shown in Figure 17. It requires the following materials:

1. steel block 1-1/8" X 1-1/4" X 1-7/8" for base
2. steel rod 1/4" X 7" for pointer
3. pressure pin 3/16" X 3/8" long, to prevent the screw from cutting into rod
4. allen set screw 1/4" X 1/4", 20 threads, to lock the pointer in position

This tool can be used for centering the knurling tools, also turning, boring, threading and cutting-off tools, or any other centering to be done on the lathe. The block should be faced on the bottom, and finished with a depression in the center, cut 1/32" deep, and 1" in diameter. This permits the base to set firmly on the flat way of the lathe bed and prevents rocking.

When the machining on the block and rod is done, including the bending, set the block on the flat way. Push the rod down to the bottom of the hole. Position the flat end of the arm with the tip of the tailstock center. Measure the distance from the top of the flat end to the point of the center. Then cut off the other end so that the flat end of rod will be about 1/8" below center. This should allow movement for adjustments.

This tool can be centered by bringing the top end or sides of the pointer against the tail stock center point, and raised or lowered in the block until the top is the same height as the tip of the tail center.

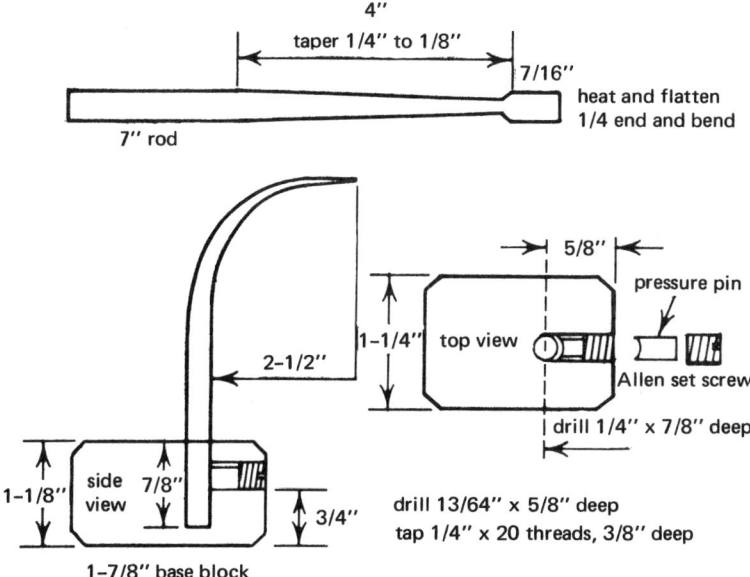

Figure 17

A final test can be made by placing a small piece of stock in the chuck and facing it with a right-hand side cutter bit. If the setting is correct the center will be cut clean, and there will be no tip left.

The cutting tool can be raised or lowered until it cuts properly, and then the point of the centering tool can be placed against the tip of the cutting bit and adjusted so the top of the points are even.

HOW TO USE THE CUTTING-OFF OR PARTING TOOL

The cutting-off operation, like knurling, is another lathe job that is considered very simple, but it can give trouble too. For instance some kinds of steel collect on the points and interfere with cutting; a hacksaw may work better. Other difficulties, and how to avoid them, will be mentioned from here on.

The point of the tool should always be set exactly on center or a trifle under, never above. If it is set too high it makes cutting difficult, requiring more pressure or making it impossible to feed into the work. If set too low it has a tendency to hook into the metal and stall the machine or break off the point.

Like knurling this job is more troublesome on small lathes than large ones, which are not as rigid. Lubrication is very important; the work should be flooded with oil at all times. If the work is allowed to run dry the friction will generate enough heat to foul up the cutting and stall the machine or break the tool. Oil is not needed for cutting cast iron. For occasional cut-off jobs, use machine or black thread-cutting oil, thinned with kerosene and applied with a squirt can. There are also the old water soluble oil and new types of cutting oils.

If there is much cut-off work to be done it is best to install a motor-driven coolant pump or rig up a can with a length of small copper tubing and a pet-cock to regulate the flow; hang it above the lathe. This will provide better lubrication and also act as a coolant.

The bottom end of the tube can be fastened so that the oil solution will drip or flow directly over the point of the tool, and be turned on before the cutting is started. The cutting can be done much faster with a steady stream flowing over the work.

When setting up these tools they should be centered first and the screw tightened a little. Then slide the carriage so that the end of the blade comes against the end of one of the chuck jaws. Adjust by tapping the shank end of tool until it is in a straight line with the end of the jaw; the blade must be at a right angle to the work. Then tighten the screw securely.

HOW TO GRIND CUTTING-OFF OR PARTING TOOLS

See Figure 18.

These tools usually come ground as shown in Figure 19 and will scrape about as much as they will cut. When ground as shown in Figure 20, the curved top of the end gives the tool a sharper cutting point and the chips curl up into coils, something like a watch spring; there is less chance for the chips to foul up the work. The front end of the cutter

Tools for Knurling / 69

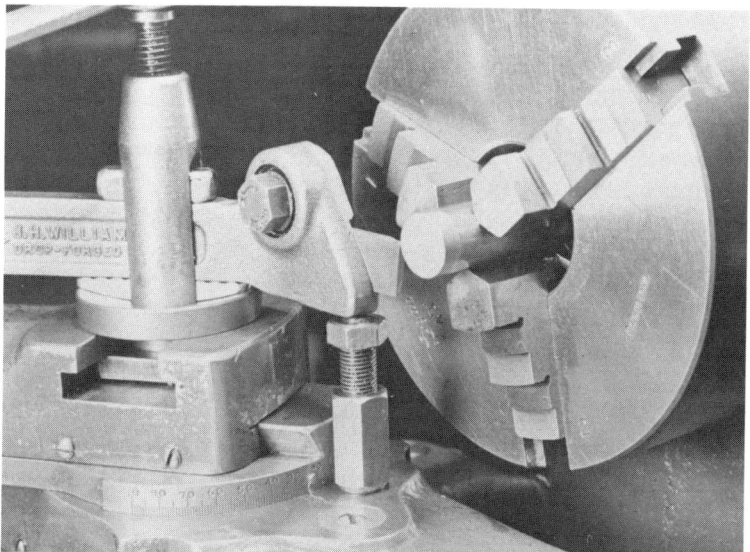

Figure 18

blade may be ground to an angle of 10° to 15°. This should result in a clean-cutting, easier operation, and the point should not generate as much heat. The side taper of these blades should provide enough side clearance.

The grinding should be done on a fine grit wheel and the point honed on top and the front end with a fine oil stone as often as necessary to keep the point sharp. If the tool is fed too fast into the work there is a tendency for the tip to dip down, and the work to climb on top of it. To help prevent this, the cross feed and compound rest gibs should be adjusted to eliminate excessive play. When the tool is positioned for the cut it may be a good idea to lock the carriage.

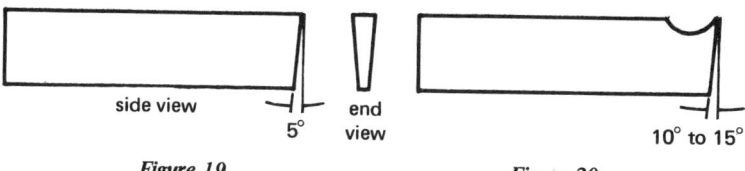

Figure 19 *Figure 20*

HOW TO MAKE A SMALL JACK FOR CUT-OFF WORK ON 9" LATHES

A small jack will be very helpful in making the setup more rigid; it can be made similar to the one outlined in Figures 21 and 22. Screw the compound rest back far enough to provide a footing for the jack. Slide the tool holder back into the tool post as far as practical; line it up and center it. Then place the jack on the compound rest base, and under the end of tool holder as shown in Figure 18.

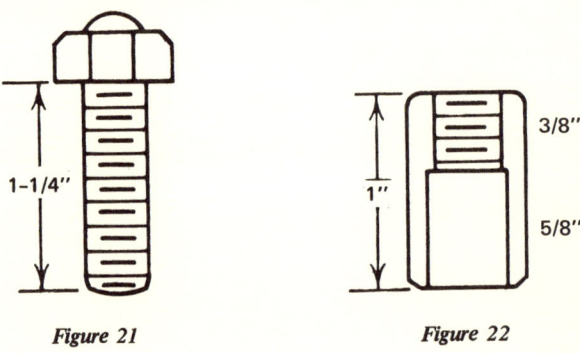

Figure 21   *Figure 22*

Unscrew the screw with your fingers until it is good and tight. This is done to make the tool more solid and prevent it from dipping down. When practical the cutting-off should be done close to the chuck. If not the work should be supported on tail center.

If this tool is set above center it will be hard to finish the cut at the center. If set too low, there will be an uncut tip left at the center. When set exactly on center there will be no tip left.

Figure 21 shows a 3/8" X 1-1/4" cap screw, 24 threads. Machine head so that it will have a rounded high center so that it can be turned more easily with the fingers. This tool would have to be made especially for different types and sizes of lathes.

Drill 21/64" hole top end last; thread with 3/8" tap 24 threads (Figure 22). This part can be made of 5/8" hexagon steel; round stock may do if it is knurled. First drill bottom part of hole with a 25/64" drill 5/8" deep, so that the screw will have clearance.

Tools for Knurling / 71

When cutting hollow pieces for bushings or washers, with a square pointed tool (Figure 23), there is usually a thin strip left on the edge of hole after the piece is cut off. To avoid this the end of the tool can be ground as illustrated in Figure 24, at an angle of about 10°. This will allow the piece to cut off clean, and when the tool is advanced a little more it will trim off the part in the chuck.

One-quarter cutter bits can be ground with a thin section about 1/16" thick and back about 1/2" from the point, for small diameter work as shown in the figures below. After the tools are centered they can be moved against the flat end of the chuck jaws and lined up straight so that they will be square with the work.

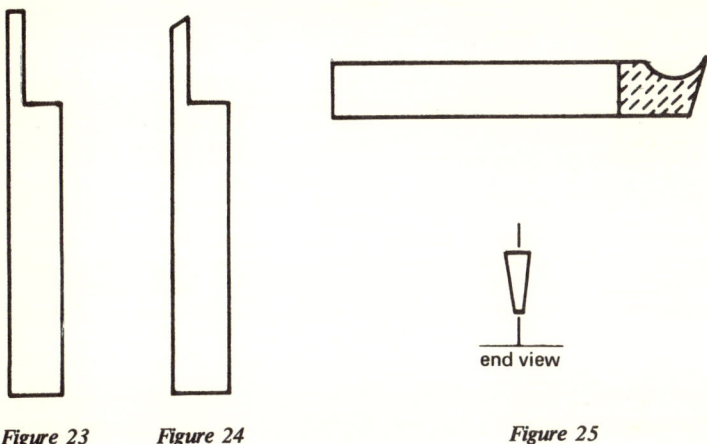

Figure 23   *Figure 24*   *Figure 25*

HOW TO MAKE SMALL SUBSTITUTE GEARS

Spur Gears

Small spur gears can be made with straight-tooth knurls (Figure 26). A large variety of diameters and widths can be made with the regular knurling process, from the smallest diameters the knurls will work on to several inches. The variety is limited to the number and pitches of the knurls available. In order to make these gears mesh properly, the same pitches would have to be used to form the teeth on all the gears that are to work together.

Figure 26

Helical Gears

Helical gears can be made with diagonal knurls, either right or left helix angles. Only one knurl is to be used. The right-hand diagonal knurl will produce the left-hand spiral, or helix teeth on the blank, and the left-hand diagonal knurl will form the right-hand helix.

Knurls are made for a different purpose than gears. The V form of the teeth is not the proper form to be used for "working gears," because the teeth rolled on the blanks with knurls would be a duplicate of the knurl teeth; they would not work well as gears.

Gear teeth are cut deeper, the roots rounded or flat. The teeth are thicker and the tips rounded or flat, and the pressure points curved. These gear substitutes should not be expected to be suitable for heavy work loads; they could be useful for temporary work such as testing out certain experimental projects. They could be case hardened after they are finished to make them last longer.

In order to have the helical gears mesh properly, they would have to be made with a pair of diagonal knurls of the same pitch; the drive gear would have to be rolled with either the right- or left-hand spiral knurl, and the driven would have to be rolled with the opposite spiral.

These gears can be made by selecting the diameters and pitches desired. Then do the knurling. Place in chuck and drill the hole to the proper size and depth. Then set up the cut-off tool; measure the width and cut it off. If hubs are needed on one or both sides, this tool can be used to cut in to get the diameter. Then withdraw it and move the carriage to get the length of the hub. Withdraw and set in the cut-off position so that the hub will be the right length. A number of gears the same size can be cut by using the longitudinal feed and knurling to the desired length.

Appendix A

Tables for Working Out Formulas

TABLE 1: 3.1416

The figure 3.1416 is frequently used in the process of working out knurling formulas and tables of diameters. This table can be used in subtraction, multiplication, and division.

Table 1: 3.1416 Multiplied

1 X =	3.1416
2 X =	6.2832
3 X =	9.4248
4 X =	12.5664
5 X =	15.7080
6 X =	18.8496
7 X =	21.9912
8 X =	25.1328
9 X =	28.2744

TABLE 2: CIRCULAR PITCH

This table is useful, too, because the number of teeth times the circular pitch of the knurls gives the circumference, which must be divided by 3.1416 to get the correct blank diameters to work on. Of course, you would have to work out these tables according to the pitches of the knurls you are going to use.

Circular pitches are for straight-tooth knurling, and transverse circular pitches for diagonal knurling. The circular pitches would have to be changed occasionally due to the effect of the arc. The decimal points in both tables may be omitted later on in some of this work. Also 3.1416 need not be written down after a few examples have been worked out in some of the formulas.

Table 2A: .064 Multiplied

Knurl: 16-TPI straight-tooth. *Circular pitch:* .064 for diameters from .328 to .671.

```
1 X =  .064
2 X =  .128
3 X =  .192
4 X =  .256
5 X =  .320
6 X =  .384
7 X =  .448
8 X =  .512
9 X =  .576
```

Table 2B: .0638 Multiplied

Knurl: 16-TPI straight-tooth. *Circular pitch:* .0638 for diameters from .671 to 2.031.

```
1 X =  .0638
2 X =  .1276
3 X =  .1914
4 X =  .2552
5 X =  .3190
6 X =  .3828
7 X =  .4466
8 X =  .5104
9 X =  .5742
```

These tables can be used to save time and perhaps some mistakes, when using Formula 12 to figure out tables of work blank diameters, processing knurls, and if you want to knurl smaller or larger diameters than have been worked out in any of the tables of diameters.

Table 3: .86603 Multiplied

1 X =	.86603
2 X =	1.73206
3 X =	2.59809
4 X =	3.46412
5 X =	4.33015
6 X =	5.19618
7 X =	6.06221
8 X =	6.92824
9 X =	7.79427

VARIETY OF PATTERNS

With the multiple pitch system of knurling it is possible to produce the following patterns and even more, with the standard knurls listed in Table 4. The patterns can be straight, right or left diagonal, and the diamond patterns can be made with right-and left-hand diagonal knurls used in pairs. Diametral pitch knurls can also be used with this system.

Table 4: Multiple Pitch Patterns

Normal TPI of Knurls	Patterns			
12	24	36	48	60
16	32	48	64	80
19	38	57	76	
20	40	60	80	
24	48	72		
25	50	75		
29	58	87		
30	60	90		
35	70			
40	80			

Appendix B

Tables of Diameters

HOW TO USE THE TABLES OF DIAMETERS

First measure the knurls you are going to use, as described in formulas 2 and 5. Then check the description of knurls at the head of the tables of diameters. If the measurements correspond, the knurls should come close to tracking on the diameters listed.

If not try a little extra pressure. If the tracking is still off, check to see if the teeth are stepping over. If so, the diameters are too small for the pitch of the knurls. Add .002 or .003 and check again. Test until the tracking is good; then add the amount you found to the diameters.

If the teeth step under, reduce the diameter until the teeth will track. Sometimes it may be necessary to add to or subtract as much as .002 to .006; even so the tables may still be useful. The knurls may track on soft metals but may need the diameters changed to correspond to the hardness of the steel.

When the tracking is once established it is just a matter of selecting the diameter in the table nearest to the one you want to work on, then adding or subtracting the amount found by the test. For example, if you had to increase the diameters .003 to get good tracking, then you could add .003 to any of the diameters listed in the table and machine the stock to that diameter.

One discouraging feature in working out these tables is that you can have the diameter worked out so that the knurls track perfectly, and a few days later you start to do a knurling job and find the knurls are out of track, even though the same setup is used. This is due to the many difficulties that are relevant to the knurling process, which in effect makes knurling one of the most exasperating of all lathe operations. Thread-cutting is easy in comparison, and it is apparent that it is regarded as a very simple operation, because it is not even mentioned in many of the books on machine-shop work, and even in the best of them the information is very limited. That is why this book was written.

The circular pitches and the transverse circular pitches at the right of the diameters listed in the tables will differ from the circular pitches of the knurls shown at the head of the tables. One reason for this is that the curve of the arc changes with the different diameters. This effects the circular pitches enough so that they must be changed at certain intervals.

The tables of diameters were worked out with formula 3 for straight knurling, and formula 6 for diagonal knurling. They will take care of the effect of the arc so that the diameters will come out somewhere near right, and the circular pitches and transverse circular pitches are more accurate for the diameters.

The other reason is that the circular pitches of the straight-tooth knurls shown at the head of the tables were obtained with formula 2, and the transverse circular pitches of the diagonal knurls were obtained with formula 5, which are close enough for some work, but are not accurate enough when working out tables of diameters.

According to tests made, all knurls from 1/2 to 3/4 diameters should work on the diameters listed in the tables of diameters, provided the pitches are the same as shown at the head of the tables. A few thousandths may have to be added or subtracted from the diameters to get good tracking.

STRAIGHT-TOOTH KNURLING

Table 5

Knurl: 16-TPI straight-tooth. *Diameter:* .628. *Number of teeth:* 31. *Circular pitch of knurl:* .063643. *Increment:* .020258. *Variation:* .020 to .021; use .020.

NUMBER OF TEETH TO BE KNURLED ON CIRCUMFERENCE					
	Blank Diameters				
11	.228	41	.832	71	1.441
12	.248–CP .065	42	.852	72	1.462
13	.267	43	.873	73	1.482
14	.287	44	.893	74	1.502
15	.307–CP .0644	45	.914	75	1.522
16	.328	46	.934	76	1.543
17	.347	47	.954	77	1.563
18	.367	48	.975	78	1.583
19	.387	59	.995	79	1.603
20	.407	50	1.015	80	1.624
21	.427	51	1.036	81	1.644
22	.447	52	1.056	82	1.665
23	.468–CP .064	53	1.077	83	1.685–CP .0638
24	.488	54	1.097	84	1.705
25	.509	55	1.117	85	1.725
26	.529	56	1.137	86	1.746
27	.549	57	1.157	87	1.767
28	.570	58	1.177	88	1.787
29	.591	59	1.198	89	1.807
30	.610	60	1.218	90	1.827
31	.630	61	1.239	91	1.847
32	.651	62	1.259	92	1.868
33	.671–CP .0639	63	1.279–CP .0638	93	1.889
34	.691	64	1.299	94	1.909
35	.711	65	1.319	95	1.929
36	.732	66	1.339	96	1.950
37	.753	67	1.360	97	1.970
38	.772–CP .0638	68	1.380	98	1.990
39	.792	69	1.400	99	2.010
40	.812	70	1.421	100	2.031

Table 6

Knurl: 21-pitch straight-tooth. **Diameter:** .621. **Number of teeth:** 42. **Circular pitch of knurl:** .04645. **Increment:** .01479. **Variation:** .014 to .015; use .015.

NUMBER OF TEETH TO BE KNURLED ON CIRCUMFERENCE

Blank diameters					
11	.163	46	.675	81	1.190
12	.178	47	.690	82	1.205
13	.192–CP .0465	48	.705–CP .04614	83	1.220
14	.207	49	.719	84	1.235
15	.221	50	.733	85	1.249
16	.235	51	.747	86	1.263
17	.250	52	.762	87	1.278
18	.264–CP .0461	53	.777	88	1.293–CP .04615
19	.279	54	.792	89	1.308
20	.294	55	.806	90	1.322
21	.307	56	.821	91	1.337
22	.322	57	.836	92	1.351
23	.337	58	.851	93	1.366
24	.351	59	.866	94	1.380
25	.365	60	.881	95	1.395
26	.381	61	.895	96	1.410
27	.395	62	.910	97	1.425
28	.410	63	.925	98	1.440
29	.425	64	.939	99	1.455
30	.440	65	.954	100	1.469
31	.454	66	.968	101	1.484
32	.469	67	.983	102	1.498
33	.484–CP .0461	68	.998	103	1.513
34	.498	69	1.012	104	1.528
35	.513	70	1.027	105	1.543
36	.528	71	1.042	106	1.557
37	.542	72	1.057	107	1.572
38	.557	73	1.072	108	1.587
39	.572	74	1.086	109	1.601
40	.586	75	1.101	110	1.616
41	.601	76	1.115		
42	.615	77	1.129		
43	.630	78	1.145		
44	.645	79	1.160		
45	.660	80	1.175		

Table 7

Knurl: 40-TPI. **Diameter:** .621. **Number of teeth:** 78. **Circular pitch of knurl:** .02501. **Increment:** .00796. **Variation:** .007 to .008. use .008.

NUMBER OF TEETH TO BE KNURLED ON CIRCUMFERENCE					
Blank diameters						
16	.129	56	.449	96	.769
17	.137	57	.457	97	.777
18	.145–CP .02531	58	.465	98	.785
19	.153	59	.473	99	.793
20	.161	60	.481	100	.801
21	.169	61	.489	101	.809
22	.177	62	.497	102	.817
23	.185	63	.505	103	.825
24	.193	64	.513	104	.833
25	.201	65	.521	105	.841
26	.209	66	.529	106	.849
27	.217	67	.537	107	.857
28	.225	68	.545–CP .02517	108	.865
29	.233	69	.553	109	.873
30	.241	70	.561	110	.881
31	.249	71	.569	111	.889
32	.257	72	.577	112	.897
33	.265	73	.585	113	.905
34	.273	74	.593	114	.913
35	.281	75	.601	115	.921
36	.289	76	.609	116	.929–CP .02516
37	.297	77	.617	117	.937
38	.305	78	.625	118	.945
39	.313	79	.633	119	.953
40	.321	80	.641	120	.961
41	.330	81	.649	121	.969
42	.338	82	.657	122	.977
43	.346–CP .02529	83	.665	123	.985
44	.354	84	.673	124	.993
45	.362	85	.681	125	1.001
46	.370	86	.689	126	1.009
47	.378	87	.697	127	1.017
48	.386	88	.705	128	1.025
49	.394	89	.713	129	1.033
50	.402	90	.721	130	1.041
51	.409	91	.729		
52	.418	92	.737		
53	.425–CP .02517	93	.745–CP .02516		
54	.433	94	.753		
55	.441	95	.761		

DIAGONAL AND DIAMOND KNURLING

Table 8

Knurls: 14-Pitch, Type 2. Diameter: .622. Number of teeth: 28. Transverse circular pitch of knurls: .06988. Increment: .02221. Variation: .022 to .023.

NUMBER OF TEETH TO BE KNURLED ON CIRCUMFERENCE					
\multicolumn{2}{l}{Blank Diameters}				
11	.249	41	.917	71	1.584
12	.271	42	.940	72	1.607
13	.294–.0706 TCP	43	.961–.0703 TCP	73	1.629
14	.316	44	.983	74	1.651
15	.338	45	1.006	75	1.673
16	.360	46	1.029	76	1.696
17	.382	47	1.051	77	1.718
18	.404	48	1.074	78	1.740
19	.426	49	1.096	79	1.762
20	.448	50	1.119	80	1.784
21	.471	51	1.141	81	1.806
22	.494	52	1.164	82	1.829
23	.516–.0705 TCP	53	1.186	83	1.851
24	.538	54	1.208	84	1.873
25	.561	55	1.231	85	1.895
26	.583	56	1.252	86	1.918
27	.606	57	1.273	87	1.940
28	.628	58	1.296–.0702 TCP	88	1.962–.07006 TCP
29	.651	59	1.318	89	1.985
30	.673	60	1.340	90	2.007
31	.695–.0704 TCP	61	1.361		
32	.717	62	1.383		
33	.740	63	1.406		
34	.762	64	1.428		
35	.785	65	1.450		
36	.806	66	1.472		
37	.828	67	1.494		
38	.851	68	1.516		
39	.874	69	1.539		
40	.896	70	1.561		

Table 9

Knurls: 20-TPI. **Diameter:** .625. **Number of teeth:** 34. **Transverse circular pitch of knurls:** .05775. **Increment:** .018382. **Variation:** .018 to .019; use .018

NUMBER OF TEETH TO BE KNURLED ON CIRCUMFERENCE					
Blank Diameters						
6	.114	41	.760	76	1.401
7	.133	42	.778	77	1.419
8	.152–.05969 TCP	43	.796–.05815 TCP	78	1.438
9	.170	44	.814	79	1.456
10	.189	45	.832	80	1.475
11	.208	46	.850	81	1.494
12	.227	47	.869	82	1.513
13	.245–.0592 TCP	48	.888	83	1.532–.05792 TCP
14	.264	49	.906	84	1.551
15	.282	50	.924	85	1.568
16	.300	51	.943	86	1.585
17	.319	52	.961	87	1.604
18	.338–.0589 TCP	53	.980	88	1.622
19	.356	54	.998	89	1.641
20	.375	55	1.016	90	1.659
21	.393	56	1.032	91	1.678
22	.412	57	1.050	92	1.696
23	.430	58	1.069–.05792 TCP	93	1.714
24	.449	59	1.088	94	1.733
25	.467	60	1.106	95	1.752
26	.486	61	1.125	96	1.770
27	.504	62	1.143	97	1.788
28	.522–.05868 TCP	63	1.161	98	1.807
29	.540	64	1.180	99	1.825
30	.558	65	1.198	100	1.844
31	.576	66	1.217	101	1.862
32	.594	67	1.235	102	1.880
33	.613–.05832 TCP	68	1.254	103	1.899
34	.632	69	1.272	104	1.917
35	.650	70	1.290	105	1.936
36	.668	71	1.309	106	1.954
37	.687	72	1.327	107	1.972
38	.706	73	1.346	108	1.991–.05792 TCP
39	.724	74	1.364	109	2.009
40	.742	75	1.383	110	2.028

Table 10

Knurls: 21-Pitch Type 2. *Diameter:* .624, *Number of teeth:* 42. *Transverse circular pitch of knurls:* .046675. *Increment:* .014857. *Variation:* .014 to .015; use .015.

NUMBER OF TEETH TO BE KNURLED ON CIRCUMFERENCE					
	Blank Diameters				
11	.167	46	.691	81	1.210
12	.182	47	.706	82	1.225
13	.197–.0476 TCP	48	.721	83	1.240–.04694 TCP
14	.211	49	.736	84	1.255
15	.227	50	.750	85	1.270
16	.243	51	.765	86	1.285
17	.258	52	.779	87	1.300
18	.273–.0476 TCP	53	.794–.0471 TCP	88	1.315–.04693 TCP
19	.288	54	.809	89	1.330
20	.303	55	.824	90	1.345
21	.317	56	.838	91	1.359
22	.332	57	.853	92	1.374
23	.347–.0475 TCP	58	.868	93	1.389
24	.361	59	.883	94	1.404
25	.376	60	.898	95	1.419
26	.391	61	.912	96	1.433
27	.406	62	.927	97	1.448
28	.421–.04723 TCP	63	.942–.047 TCP	98	1.463–.04692 TCP
29	.436	64	.957	99	1.478
30	.451	65	.972	100	1.493
31	.467	66	.987	101	1.508
32	.481	67	1.002	102	1.523
33	.496	68	1.017	103	1.538–.04691 TCP
34	.511	69	1.032	104	1.553
35	.526	70	1.047	105	1.568
36	.541	71	1.062		
37	.556	72	1.077		
38	.571–.0472 TCP	73	1.092		
39	.586	74	1.107		
40	.601	75	1.122		
41	.616	76	1.137		
42	.631	77	1.152		
43	.646	78	1.167		
44	.661	79	1.182		
45	.676	80	1.197		

Table 11

Knurls: 25-TPI. *Diameter:* .618. *Number of teeth:* 42. *Transverse circular pitch of knurls:* .04623. *Increment:* .014714. *Variation:* .014 to .015; use .015.

NUMBER OF TEETH TO BE KNURLED ON CIRCUMFERENCE					
	Blank diameters				
16	.240	51	.757	86	1.274
17	.255	52	.772	87	1.289
18	.270–.04728 TCP	53	.787	88	1.303
19	.285	54	.802	89	1.318
20	.300	55	.817	90	1.333
21	.314	56	.831	91	1.347
22	.329	57	.845	92	1.361
23	.344–.04699 TCP	58	.860–.0466 TCP	93	1.375–.04646
24	.359	59	.875	94	1.390
25	.374	60	.890	95	1.405
26	.388	61	.904	96	1.420
27	.402	62	.918	97	1.435
28	.417–.04678 TCP	63	.933	98	1.450
29	.432	64	.948	99	1.464
30	.447	65	.963	100	1.479
31	.462	66	.978	101	1.494
32	.476	67	.993	102	1.508
33	.491	68	1.008–.04656	103	1.523–.04644
34	.506	69	1.023	104	1.538
35	.521	70	1.038	105	1.553
36	.535	71	1.053	106	1.568
37	.549	72	1.068	107	1.583
38	.564–.04667	73	1.082	108	1.597
39	.579	74	1.097	109	1.611
40	.594	75	1.112	110	1.626
41	.609	76	1.127	111	1.641
42	.623	77	1.142	112	1.656
43	.638	78	1.156	113	1.670–.04644
44	.653	79	1.171	114	1.685
45	.668	80	1.185	115	1.700
46	.683	81	1.200	116	1.714
47	.698	82	1.214	117	1.729
48	.713	83	1.229–.04653 TCP	118	1.743–.04642
49	.728	84	1.244	119	1.758
50	.742	85	1.259	120	1.773

Table 12

Knurls: 30-TPI. *Diameter:* .612. *Number of teeth:* 50. *Transverse circular pitch of knurls:* .038453. *Increment:* .012239. *Variation:* .012 to .013; use .012.

NUMBER OF TEETH TO BE KNURLED ON CIRCUMFERENCE					
	Blank diameters				
16	.200–.03927 TCP	51	.629	86	1.060
17	.212	52	.641	87	1.072
18	.224–.0391	53	.654	88	1.084
19	.236	54	.666	89	1.096
20	.248	55	.678	90	1.109
21	.260	56	.690	91	1.121
22	.272	57	.702	92	1.133
23	.284–.0388	58	.714–.0387	93	1.145–.0387
24	.296	59	.727	94	1.158
25	.309	60	.739	95	1.170
26	.322	61	.751	96	1.182
27	.334	62	.764	97	1.195
28	.346	63	.776–.0387	98	1.207–.03867
29	.358	64	.788	99	1.219
30	.370	65	.801	100	1.231
31	.383	66	.813	101	1.243
32	.395	67	.825	102	1.255
33	.408	68	.838	103	1.268
34	.420	69	.850	104	1.280
35	.432	70	.862	105	1.292
36	.445	71	.875	106	1.305
37	.457	72	.887	107	1.317
38	.470	73	.899–.0387	108	1.329
39	.482	74	.911	109	1.342
40	.494	75	.924	110	1.354
41	.506	76	.936	111	1.366
42	.518	77	.948	112	1.379
43	.530–.0388	78	.961	113	1.391–.03867
44	.542	79	.973	114	1.403
45	.555	80	.985	115	1.415
46	.567	81	.998	116	1.428
47	.580	82	1.010	117	1.440
48	.592–.03875	83	1.022	118	1.452
49	.604	84	1.035	119	1.465
50	.617	85	1.048	120	1.477